# ESCAPING THE

# SOFTWARE

# MONEY

# PIT

## A Winning Roadmap for Managers and Executives

## MATT COOK

McKenna-Heyl Press

Printed in the USA

First Edition October 2013

McKenna-Heyl Press

Cover design by Meredith McKenna Cook

ISBN: 0615820840
ISBN 13: 9780615820842

# Table of Contents

| | | |
|---|---|---:|
| **INTRODUCTION** | | **1** |
| ONE | **THE MINEFIELD** | **5** |
| | What Is Software? | 7 |
| | Fuzzy Marketing | 8 |
| | The Self-Perpetuating Software Market | 11 |
| | Ridiculously Long Projects | 13 |
| | Why Does It Cost So Much? | 16 |
| | Why So Poor A Track Record? | 18 |
| | Academic Studies | 20 |
| TWO | **MONEY PIT CASE STUDIES** | **25** |
| | **Case Study #1: Complacency Fuels A Disaster** | **25** |
| | Case Study #2: Nike's Debacle With Supply Chain Software | 29 |
| | Case Study #3: Denver Airport Baggage System | 33 |
| | Case Study #4: Did We Get Anything Out Of This? | 35 |
| | Case Study #5: The 2010 Census | 38 |
| | Case Study #6: Customization Surprises | 41 |
| THREE | **WINNING CASE STUDIES** | **45** |
| | Case Study #7: Conversion To SAP In 90 Days | 45 |
| | Case Study #8: A Win On Dashboard Software | 49 |
| | Case Study #9: SaaS Success Story #1 | 52 |
| | Case Study #10: SaaS Success Story #2 | 55 |
| FOUR | **CASE STUDY SUMMARY** | **59** |
| | Losers | 59 |
| | Winners | 60 |

FIVE **THE HUMAN FACTOR** 63

SIX **NIP THE LOSERS IN THE BUD** 73
Good Reasons To Invest 74
Bad Reasons To Invest 76
The Shiny Object 77
You As Gatekeeper 79
Evaluate The Assumptions 81
Connect The Dots 82

SEVEN **STEP FIRMLY INTO THE FUTURE** 83
Keep The Scope Manageable 84
Draw The Future In Detail 85
Key Steps In Defining The Future 87

EIGHT **HAVE A STRATEGY!** 91
What's A Strategy? 91
Make The Strategy Fit Your Business 92
Tread Carefully With Custom Software 94
Know What Your Choices Are 96
Be Smart About Integration 100
Have A Structure For Evaluating Your Choices 103

NINE **UNDERSTAND YOUR OPTIONS** 107
ERP Software 108
Big ERP 110
Other ERP 111
Manufacturing 113
Supply Chain 116
Finance 118
CRM and Sales Management 120
Analytics 121
Big Data 123
Data Virtualization 125
Enterprise Information Management 126

TEN **CONSIDER CLOUD, SAAS AND MOBILE** 129
Cloud Computing 130
Software-as-a-Service 131
Mobile Computing 134

ELEVEN **MANAGE RUTHLESSLY** 141
Create A Ruthless Team 142
Embrace Structure 146
Make People Pay Attention 155
Kill The Project Killers 157
Learn From Your Own Experience 159

TWELVE **SURF THE TSUNAMI** **161**
Pay Attention To The World Around You — 162
Your Employees Will Dictate Your Software Choices — 162
Traditional ERP Continues To Look Unnecessary — 163
Better Options Replace Traditional Data Warehousing — 164
In-Memory Computing — 166
The Visualization Of Data — 167
A Large Menu Of Choices — 168
Muddy Waters In The Software Market — 169
The Elusive Breakthrough — 171

THIRTEEN **THE ESSENTIALS TO WINNING –
A SUMMARY** **173**
Understand The Minefield — 173
Nip The Losers In The Bud — 176
Step Firmly Into The Future — 177
Have A Strategy! — 180
Be Smart About Integration — 182
Have A Structure For Evaluating Your Choices — 183
Understand Basic Software Choices — 184
Take Advantage Of SaaS, Cloud And Mobile — 185
Manage Ruthlessly — 186

APPENDIX **191**
Ten Easy Ways To Lose — 191
Lingo Translator — 194
What's An ROI? — 205
How Do I Get My ROI? — 205

# Introduction

"Confidence: that feeling you have just before you fully understand the situation."

– Oft-used quote reportedly first seen in print in 1961, barrypopik.com

In the 21st century, many enterprises proceed with new software systems and soon find themselves in a money pit. Software projects are failure-prone and expensive. The software industry's marketing is confusing. It's hard to determine how a proposed system will produce financial benefits, to validate claims by software firms, and to compare one software solution against others.

This doesn't have to be. And you shouldn't have to become an IT expert and project management guru to turn a profit on your investment – that's why I wrote this book, a plain-spoken guide to smart and winning software investments, for managers and executives in all types of organizations: business, government, education and non-profit.

Throughout this book I use the terms "organization," "enterprise," and "company" interchangeably, and by "enterprise software" I mean any software used to enable one or more processes within an organization, whether it's software just for the sales team or software to manage multiple functions like accounting, purchasing, manufacturing, inventory and payroll.

To launch an enterprise software initiative is to enter a minefield.

Company X decides to invest $5 million in new software. Will the firm succeed? Will it really cost "only" $5 million and deliver the financial returns expected? Studies say it's unlikely, and the landscape is littered with notable failures.

Examples:

- In September 1999 Hershey Foods CEO Kenneth Wolfe told analysts that the company was having trouble with its new enterprise software and as a result would not be able to keep up with order demand in the upcoming fall candy season. Hershey's stock dropped by 8%, and the company struggled with an estimated $100 million in lost sales that year.

- In 2008 the North Carolina Department of Health and Human Services contracted with Computer Sciences Corp. (CSC) to rewrite the state's Medicaid claim processing software. The project is expected to be completed in the summer of 2013, two years late and at a cost of nearly double the original contract price of $265 million.

A well-planned and smartly budgeted software investment becomes a vast hole sucking money, time, worry, family life and peoples' careers into its dark abyss. But there are many things enterprises can do to dodge the money pit and to instead manage really smart and successful software investments – that is what I give you in this book, based on 17 years managing software projects large and small, as well as case studies of both failed and successful projects.

It seems companies are still making the same mistakes with software that were made 20 years ago. I've seen the bad decisions firsthand – lack of understanding of what solutions are available in the market, unclear business goals, poor business case assumptions or no business case at all, over-reliance on consultants, unrealistic timelines, poor team and project manager selection, and lack of project risk management.

My own experience matches what studies have found: that a large percentage of projects fail; meaning they take way longer or cost way more than planned, are stopped or abandoned entirely, or are eventually completed but along the way resulted in enormous damage to the business, as in the colossal failure of Denver Airport's new automated baggage handling system in 1995 that delayed the new airport's opening by 16 months and cost $560 million more than planned.

Enterprises can't just stop investing in software for their operations. These applications make everything run efficiently, and new business needs often require new or updated software.

This book is relevant for an increasing number of managers and executives whose worlds are becoming digitized and "application-ized." If you're not careful you'll get sucked into the details. Stay above the details and remember that no matter how awesome technology and software become, your enterprise still has a job to do, whether it's educating 40,000 students, making airplane parts or processing potatoes. The software you buy isn't going to make that job any easier or more effective; it's what you do with it that will make a difference.

# 1

# THE MINEFIELD

"Those parts of the system that you can hit with a hammer (not advised) are called hardware; those program instructions that you can only curse at are called software."

> – Levitating Trains and Kamikaze Genes: Technological Literacy for the 1990's, by Richard P. Brennan

The F-35 Joint Strike Fighter, first conceived in the mid-1990s, is a multi-purpose fighter jet that evades radar and anti-aircraft missiles, can fly at 1,200 mph, and can land vertically. While over 40 planes have been manufactured, the program is way behind schedule and the program's costs have grown from an initial estimate of $177 billion to over $1 trillion, according to a *Wall Street Journal* article authored by defense industry analysts Arthur Herman and John Scott. Herman and Scott say the Department of Defense (DOD) has identified *software development* (the plane's systems run to nearly 10 million lines of source code) as one key issue behind the F-35's problematic history, noting that the software needed for producing the plane won't be completed until 2017.

Not making fighter jets with 10 million lines of code? You'll still find that a minefield awaits you regardless of your software endeavor.

Fortunately, many have gone before you into said minefield, and plenty of insights are available to those who want to launch their projects in an informed way.

This chapter explores the maddening nature of software, why it has historically been problematic to develop and use good software, and the studies that have been done of the track records of software investments.

Regarding new systems for businesses or government entities, most people think that:

- Converting to new software is simply a matter of writing or purchasing the right application, installing it and turning it on;

- Everything will go fine as long as you have the right technical people involved;

- What the new software will do for the enterprise can be defined accurately beforehand, and it's unlikely the new system will not perform as expected;

- The time it takes to implement a new system is predictable, and delays are unlikely

In short, there is a general presumption that the project will succeed. While it's OK to have that positive outlook, it is more beneficial to be aware that the odds are your project – like many that have gone before – will likely run into problems, and to be well-prepared to deal with those problems.

**To dodge the software money pit you must first understand software, its true nature, why its development takes so long and costs so much money, and why the software market is such a jungle.**

## What Is Software?

Software is like no other product on earth – it is a collection of millions of lines of instructions telling a computer what to do. You can't "see" software; reading the lines of code would tell you nothing unless you had written the code yourself, and even programmers themselves can easily forget what they did. You must *imagine* software, and are left to rely on what the software's creators say about it.

That good software is indispensable goes back to one of the first-ever software projects: an effort in the early 1950s to link together data from radars along the Eastern seaboard that were monitoring possible air and seaborne threats to the United States. Software, it was discovered, could collect, compare and plot radar data on paper much faster than human beings could.

But from software's early beginnings as an industry in the 1950s and 1960s, business managers have struggled to understand the systems they buy, and the people and firms that market them.

In his excellent history of the software industry, Martin Campbell-Kelly describes the origins of software programming in the 1950s: "Only weeks after the first prototype laboratory computers sprang to life, it became clear that programs had a life of their own – they would take weeks or months to shake down, and they would forever need improvement and modification in response to the demands of a changing environment." Some things haven't changed.

Mr. Campbell also describes what must be one of the earliest maddening software experiences. General Electric had purchased a Univac computer in 1954, but it took "nearly 2 years to get a set of basic accounting applications working satisfactorily, one entire programming group having been fired in the process. As a result, the whole area of automated business computing, and Univac especially, had become very questionable in the eyes of many businessmen."

In 1979 Gideon Gartner, formerly a top technology analyst at Oppenheimer & Co., capitalized on the emerging general confusion about IT and software and formed an advisory firm to help companies with technology decisions. Gartner Inc. now has a market value of $4.3 billion, over 5,000 employees and customers in 85 countries.

Fueling Gartner's early success was a prevailing sense of what was known then as "FUD," or fear, uncertainty and doubt, among enterprises as they began investing heavily in technology. It was tough to understand emerging technologies then, and it still is today. Even though we have software tools far superior to what was available in the 1950s and even the 1980s, FUD still exists.

## Fuzzy Marketing

Part of the FUD is fueled by the way the software industry markets its products. Some stories will illustrate this, but before going into those I need to make clear that this book is not an expose about how software firms are sneaky and evil...not at all. This section and others like it in the book are intended to prepare you for how you will be approached, in a sales and marketing sense, by those wanting to sell you system solutions. Understand that every industry markets its products – software is no different. Understanding these tactics and knowing how to deal with them is one more ingredient in making your investment a winner.

### San Francisco, 1996

I was going to try one more time to understand what the speaker, a developer for a large software firm, had just presented. We were at the firm's user's conference in California. The firm at the time sought to compete in the booming ERP market with an integrated "suite" of "best of breed" applications anchored by its flagship financial and database software, targeted toward the consumer packaged goods (CPG) industry.

In the 1990s, with Y2K on the horizon, companies were overhauling their systems. End-to-end integration was catching on as a must-have. "Enterprise" software was hot. The market was frothy; companies were spending a lot of money; FUD was prevalent. But software people were speaking a different language than business people.

One of this firm's claims was that its new suite of software applications was "changing the way companies compete," so I asked the speaker how the software did this. He responded by describing the "emerging strategies in CPG companies of 'integrated data pools and seamless transaction integration' that conflict with legacy platforms and point solutions" and how the company would "evolve development to the core suite in sequential dot releases the foundation for key strategic functionalities." Not the answer I was looking for.

**Software marketing is clouded by corporate-speak and terms designed to push certain emotional buttons.** Never have a conversation about the software product on the vendor's terms. You must force the vendor to speak in straight business terms how the application solution is going to drive the financial returns expected; give the company examples of what you want to fix and let them tell you how they will fix it and what value your business will derive.

The developer's lingo-laden presentation is still classic software marketing today; call it preamble – a lot of build- up without revealing what exactly you're building up to.

The decision process for buying a PC is straightforward: price, features, memory, processing speed, and version of operating system are clearly stated. You know exactly what you are buying, and based on publicly available reviews, what to expect in terms of performance. To acquire IT software for a business or government enterprise is an entirely different endeavor.

**Nearly every software company today uses the following approach: present an idea; paint a picture of a future state (what if you could...), that is demonstrably better than what exists today,**

**and then introduce the software application as the path to that future state.** You, as the intended recipient of this sales pitch, must get past this quickly, and get on to the specifics.

**New York, 2012**

Recently I began exploring "cloud" software for integrating our company with its trading partners. This is how one company explains the benefits of its cloud software solution:

"Cloud supply chain platforms invert the traditional EDI (Electronic Data Interchange) hub equation by moving the data processing and linking logic from the partners at the ends of the spokes to the center hub itself. In this model, the entire value chain community leverages a common core technology utility so that all partners link to a single version of supply chain truth across the entire network."

Here you have the preamble and lots of jargon made to sound sensible (who can resist "a single version of supply chain truth across the entire network"?). In the section that is supposed to answer "what is this technology made of?" I come across this explanation:

"First, importantly, it is technology designed for the deep and detailed processes of a specific B2B domain. It is not generic. In this respect it diverges sharply from traditional EDI technologies that were designed for file transport and translation across all processes. Traditional systems take a "one size fits all" to information exchange: EDI file delivery for everyone, but a complete picture for no one."

There's more:

"In rich process domains such as supply chain execution, where the processes are long-lived and highly collaborative, what is needed is an information transformation layer that is knitted to the specific business processes at hand. Only by understanding the underlying and highly detailed data models and linkages of the critical business objects themselves can an information exchange platform create tight mappings to process. If an EDI VAN (Value Added Network) transports file packets, but is blind to the contents, the next generation platform interprets at the content level — it applies technology to the inside of the packet, in other words."

None of this made sense to me, so I met with the firm's representatives.

What they mean, as it turns out, is that their cloud software acts like a big private Facebook page, where your partners post data and transactions, as compared to traditional EDI, which is like passing a message to your partners in a straw where only you and they can see what's in the straw.

**One of the best ways to start a constructive conversation with a vendor is to say, "Ok I'm stupid. Connect the dots for me.** You can link me to my trading partners like we're all on Facebook. So what? Do I get rid of my EDI? What does that save me? Do I reduce inventories and order lead times? Will my customers and suppliers love me for doing this and will they partner with me on this approach?"

If a software vendor is good, that firm will assess your business situation first – your operation and processes – before trying to sell you anything.

## The Self-Perpetuating Software Market

Also contributing to FUD and the maddening nature of software is the way firms, consultants and advisory services interact with one another.

In the process of thoroughly confusing yourself trying to scan the software market for something resembling a solution to your problem, you will at some point realize that there aren't many places to go for an objective appraisal of applications or vendors. That's because, in my view, all the players in the software market – the firms themselves, consulting companies, advisory services, user groups, associations and conference organizations, and hardware and network providers – act in a way that is (shocking!) mostly designed to perpetuate and increase spending on software and related products and services.

Yes, firms strive to be competitive and earn your business, and will tell you why their solutions are better than others. But few players in the software market – even subscription advisory services – will solidly endorse or, alternately, steer you clear of, a particular software or consulting organization or its products and services. There's just very little clear-cut information to be had. Advisory firms don't want to dismiss anyone because they rely on all players to provide them information, buy their services, and attend their (expensive) conferences. IBM is not going to advise you that SAP or Oracle would be a terrible choice – their consulting revenues depend on a stream of projects implementing those same applications.

There are several well-respected organizations providing in-depth research and advisory services related to the entire IT industry. Their services and publications are indeed excellent, thorough and detailed, but only directionally conclusive and even then in a rather general and not exactly plain-spoken way, and accompanied by caveats. Consider this summary statement from one such respected research firm, in a publication reviewing software for analyzing product assortment: "Use DemandTec if optimizing your assortment in a collaborative retailer and manufacturer environment, the ability to leverage localized incrementality and demand transfer analytics or leveraging your shopper insights is a key priority for your company." Got it?

Adding to the noise are industry associations and a slew of online newsletters that seem to multiply over time...there's CIO Magazine, CIO Research, IT Whitepapers, IT Whitepapers Business Intelligence, Enterprise Business Alert, Business Management Alert, Mobile Alert, Networking Alert, InfoWorld, Supply Chain Technology Bulletin, BI Technology Bulletin, and on and on...a flood of information, discussion, interviews, blogs, announcements, conference invitations and webinars.

All of it contributes to a general feeling of being in the woods, having no idea what the forest looks like. But one thing is sure: every topic is urgently important and deserving of weighty consideration equal to that given the most pressing and critical questions of our time.

Now hitting your inbox: "New survey - High performance analytics: Are you ready?" and "7 Best Practices for Developing in the Hybrid Cloud." These digital marketing efforts are fueled by advertising for IT products and services – again, the self-perpetuating nature of the market.

**You'll have little choice but to listen to the marketing pitches from software firms and from firms that custom design software.** It's part of the process. But get past this stage quickly and change the conversation. Move it away from how great the solution is to whether and how it fits with the specific business case you have in mind.

Later chapters will contain a lot more detailed information about the software market and how to navigate it to find the best options.

## Ridiculously Long Projects

In the time is takes to purchase and implement a company's software systems, an infant can become a first grader. Degrees can be earned. And a lot of money can be spent. This is one more reason why people find software so maddening.

The classic ERP implementation, which many companies dove into in the 1990s, is a huge endeavor that can last years for even the most tech-savvy companies. As Cynthia Rettig wrote in an October 2007 MIT Sloan Management Review article, ERP systems are "massive programs, with millions of lines of code, thousands of installation options and countless interrelated pieces…. they introduce so many complex, difficult technical and business issues that just making it to the finish line with one's shirt on is considered a win."

More often than not, studies have found that the original scope is not achieved, and it may take years to optimize the solution so that it delivers real value.

Lengthy implementations have become a sore point for software buyers. Companies and governments often see large software projects as a nuisance. There has been a backlash against the big, interconnected, and costly-to- implement applications. Executives no longer readily accept big dedicated teams working on a project.

You may have seen the cartoonish billboards that say "Down with big ERP." The ads are part of a campaign begun in 2009 by, ironically, one of the largest ERP software companies, Infor. Sensing the negative turn against large software implementations, the company has sought to differentiate itself from its large competitors SAP and Oracle.

In launching the campaign in November 2009, Infor's then-CEO Jim Schaper said: "Obviously our industry hasn't done the greatest job serving their customers. Software implementations have become disruptive. They've become known for over-budget and seemingly never-ending implementations, increasing maintenance costs every year, and forcing customers to upgrade or exchange software when it wasn't advantageous or economical to do so. In many cases, the 'software experience' has been anything but a good experience."

Why does it take so long? Software is the result of converting information – human thought – into computer instructions. The "information" could be as simple as a mathematical formula, or as complex as the mind's sequential thoughts and logic. Much of the time is spent understanding what the user wants and how to convert it into business rules that a computer can apply. Large blocks of time are also spent testing the software code to make sure it does what it was intended to do.

Another reason that projects take so long is that the software must fit hand-in-glove to your business process. Remember, the software knows nothing about your business. It is told by its lines of code to perform certain calculations and transactions using data you provide. If you could talk to the system, a conversation about just one of many business processes might go something like this:

You: "We're ready to set up the new system to process sales orders."

Software: "OK. First thing you need to do is upload into my database all of your customers, and all of the products they buy."

You: "That'll be easy. I already have a list of customers and products ready."

Software: "I'll need to know customer names, addresses, what products they can buy, at what prices, what channel of business they are in, which sales rep gets credit for the sale, whether any commissions are to be calculated and paid to a broker, how to calculate the commissions, where to put that amount so that the broker gets paid, how you want me to pay the broker – check, EFT or other? – the customer's hours of receiving, their fax and telephone numbers, who to contact with questions, whether you ship to them or they pick up their order, whether they get any discounts and if so for which products, under what conditions, how to calculate the discount and how to categorize the discount whether it is a sales promotion, customer loyalty discount, order size discount, or something else, their credit limits, banking information, tax ID number, how their orders are to be packaged, what the lead time will be from receipt of order to shipping, which warehouse should ship their order, whether sales taxes apply to their purchases, whether customs duties are to be paid, whether I should check their order for errors, what I should check, what constitutes an error, and what to do if I do find an error, whether you want me to send an order acknowledgment back to the customer, where I should send the acknowledgment, how you want me to send it, and how the acknowledgment should look, what information it should have on it, whether you want to be able to print a copy of the sales order, acknowledgment and invoice, how you want the invoice amount to be calculated, how you want the invoice to look and where and how you want it to be sent, what account you want me to post their invoice amount to as sales revenue, how long the customer has to pay the invoice, and for products I will need…"

You: "Wait, what? We don't need half that stuff!"

Software: "Well, I do, if you want me to process your sales orders accurately."

You: "But this is going to take forever!"

Software: "Not my problem. It should only take you a few weeks to get your customer data together in an upload file."

You: "A few weeks just to load customer data?! We're supposed to have this whole project done in about six months!"

Software: "Ha! What genius came up with that? The project should take about a year, maybe more. Now get going and get me your customer and product data, and make sure it's **perfect,** because if it isn't I won't let it load into my database…"

**It's simply the nature of business software that, if it is to function correctly, it must know and execute in great detail all of the rules of your company and your business processes.** It is not unlike telling a human being what to do, for example, with each customer, in every possible situation, and when to do it. The difference is that software allows no slack. If it encounters a situation (incorrect customer number), it doesn't come and find you to ask about it, it just stops and displays an error message, letting you figure out what went wrong.

## Why Does It Cost So Much?

The short answer to why enterprise software costs so much is that implementing it takes so long (see above!), even if everything goes perfectly, which happens exactly as often as Haley's Comet passing through our skies. It's expensive for one reason: specialized, and therefore expensive, skills. It takes expensive skills to:

- write the software in the first place;
- modify it to your precise business needs; and

- install and test it and fix problems before you can use it.

**The three biggest cost buckets of a software investment are implementation, software modifications, and the cost of delays or disruption to the business.**

What is "implementation"? It is the process of making your business function using the new software, or "integrating" the software into your business, however you choose to look at it. Companies have different philosophies about this; some insist the software must be modified to accommodate the way the business functions; others believe in keeping the software as "vanilla" as possible by changing processes to fit the way the software was designed to work.

There is probably a happy medium, but I tend toward keeping it vanilla because of the costs involved in customization, and because unforeseen problems are more prevalent in complex solutions. I know of an SAP implementation in Brazil going on at the moment where the business has identified over 4,000 man-hours of required modifications.

"Implementation" is also the process of matching each step in your business process to corresponding steps in the software. A business "process" is usually something like "ship a customer order" or "receive a shipment from a supplier." There might be 100 or so distinct business processes in a company, each with five to eight steps or transactions involved, so a software implementation could involve matching all of those 500 to 800 steps or transactions to the new software, and that takes time, knowledge of your business, and knowledge of the new software. That's why implementations are expensive: high cost per hour multiplied by many hours.

But if a perfect project is expensive, imagine how expensive a delayed or failed project can be. Failure is the norm, according to some studies, defined as over budget, not meeting implementation dates, or not delivering functionality as expected. I would add to that list, from personal experience, failure also includes unexpected business disruption, like temporarily shutting down a manufacturing plant or shipping to your customers a day late. So the fact that software implementations are perceived to be wildly expensive is not just because software implementations are wildly expensive anyway – they also have a high failure rate, which only adds to the cost.

## Why So Poor A Track Record?

The track record for business software projects is poor because the endeavor itself is complicated, with a lot of moving parts, unpredictable outcomes, surprises, bugs and human error. For people using the new system it can be a shock. I once heard a CEO say his company's SAP implementation was like "a fish riding a bicycle."

How can it be that really smart people – people with advanced degrees, years of business or management experience, and technology expertise – formulate an investment concept, choose the solution and the business processes that go along with it, spend oodles of time and money on it, only to fail? How can it be that failure is the norm rather than the exception?

One reason is the nature of the work required. A paper published by J. Stephanie Collins and Susan Schragle-Law of Southern New Hampshire University clearly explains the unique and troublesome aspects of IT endeavors. Below is an extended excerpt; some words are highlighted for emphasis:

"Information technology (IT) projects … are challenging, in that a new IT project involves **innovations that may not have been used before**, and may include some measure of **invention and creativity**. This often requires flexibility and false-steps. In the context of project cost and time estimation, this may be very difficult to accomplish.

"IT projects **require a high level of skill**. The skills for developing or changing a technological system are complex and may span several different disciplines.

"IT projects may **be high risk, since they may be strategically important to the organization, or even important for pure survival**. IT projects tend also to be **expensive**, and thus may be high risk for the organization. Typical projects for large organizations **cost many multiples of millions**. The outcomes are stated before the task of developing an IT project begins, though the goals change as the project proceeds. As the goals change, the scope may change.

"The tasks associated with the projects are non-repetitive and non-routine in nature and involve considerable **application of knowledge, judgment, and expertise**. This makes **estimates for completion times difficult**. Most project teams are time-limited. They produce one-time outputs, such as a new product or service to be marketed by the company, or a new information system (Bikson, Cohen & Mankin, 1996)."

So the software project you are about to launch has the following inherent characteristics:

- Innovations that may not have been used before
- Invention and creativity
- Requiring a high level of skill
- High risk
- Strategically important to the business
- Important for pure survival
- Expensive
- Costing multiples of millions
- Requiring knowledge, judgment and expertise
- Difficult to estimate when it will be completed

Few people understand how hard it is to make a complex software system run flawlessly. I plug it in and it should work, right? That same attitude permeates the executive and senior manager levels, so that big dollar software projects are normally expected to succeed, on time and on budget. But most of the time they don't.

## Academic Studies

Studies over the past 15 years show failure rates ranging between one-third and two-thirds. Failure is defined in various ways - over budget, taking much longer than planned to implement, causing major business disruptions, or simply abandoned. The reasons for failure are many: a misunderstanding of the software application itself, overly complex solutions, bad project management, unrealistic goals, constantly changing requirements, and a good deal of miscommunication.

### The OASIG Study

OASIG, an organizational management group in the UK, commissioned a study in 1995 that involved interviews with 45 experts in management and business who had extensive experience with information technology projects either as consultants or researchers. The in-depth interviews revealed a dismal 20%-30% success rates for IT projects, and the reasons cited for the overall poor track record were:

- Failing to recognize the human and organizational aspects of IT;

- Weak project management; and

- Unclear identification of user requirements.

### The Chaos Report

The Standish Group is a research and advisory firm that in 1995 published The Chaos Report, which was intended to explore:

- The scope of software project failures;

- The major factors that cause software projects to fail;

- The key ingredients that can reduce project failures.

The Standish Group compiled data from 365 respondents, primarily executive IT managers. The findings:

- Only about 15% of IT projects were completed on time and on budget;
- Thirty-one percent of all projects were canceled before completion;
- Projects completed by the largest American companies had only about 42% of the originally proposed features and functions.

The firm extrapolated the results to estimate that in 1995, *80,000* projects were canceled, representing approximately $81 billion in wasted spending.

The study listed 10 reasons for success and 10 reasons for failure, with the top three in each category being:

Reasons for success

- User involvement;
- Executive management support;
- Clear statement of requirements.

Reasons for failure

- Lack of user input;
- Unclear requirements;
- Changing requirements and specifications,

### The KPMG Canada Survey

In 1997 accounting firm KPMG studied why IT projects fail. The top reasons were:

- Weak project management, including insufficient attention to risk management;
- Questionable business case for the project;

- Inadequate support and buy-in from top management.

**The Conference Board Survey**

In 2001 The Conference Board surveyed 117 companies that had started or completed ERP software projects. The results showed that:

- Forty percent of the initiatives did not produce the expected benefits within a year of completion;

- On average respondents reported  spending 25% more than expected on the implementation and 20% more on annual support costs;

- Only one-third of the respondents said they were satisfied with their results.

**The Robbins-Gioia Survey**

In 2001, management consulting firm Robbins-Gioia queried 232 companies across a range of industries about their IT investments, particularly investments in ERP systems. Of the companies that already had an ERP system in place or were in the process of implementing one:

- Fifty-one percent said the implementation of the new system was unsuccessful; and

- Forty-six percent said they believed their organization didn't know how to use the ERP system to improve business results.

One more study of software project failures is particularly useful – IAG Consulting's report "The Impact of Business Requirements on the Success of Technology Projects." The report suggests that projects can be expected to fail 68% of the time *if the business requirements phase of the project is done poorly.*

The business requirements phase describes the hard and tedious work of breaking down every single activity the software will support and documenting for each activity what exactly the software will do: calculate, display, send, store, print, determine, query, email, etc.

Most companies simply don't know how to document and manage business requirements (so they hire consultants who do). That's because *few people think in terms of 'business process,' but in matching software to your business it is absolutely essential!*

What should you take away from these studies? How can this information help you win? Two main points:

- Projects fail for many different reasons, seemingly making it hard to pinpoint the most important two or three things;

- Nearly all of the main reasons for failure can be tied back to *human factors*; said another way, the likelihood of you winning is directly correlated to the decisions you make, the strength of your project team, the way they manage the project, the way you manage the team, and particularly the strength of your project manager (PM).

By way of illustration, a strong PM and team wouldn't let some of the most common failure factors happen, such as:

- Letting requirements change frequently;
- Building an overly complex solution;
- Excluding user involvement;
- Not using top management to secure time and resources;
- Ignoring risks;
- Letting deadlines slip.

Projects often lack enough experienced people. The selected PM may have deep knowledge of the business processes concerned, but little or no project leadership capabilities. Companies don't want to dedicate their best people to these projects because they are needed by the business. It's usually a question of "who can we dedicate to the project?" not "who are the best people for the project?"

Which means in the end it is about money, because people cost money. My opinion: if you can't afford to hire the best people to manage your software investments, you can't afford software investments.

# 2
# MONEY PIT
# CASE STUDIES

"Software undergoes beta testing shortly before it's released. Beta is Latin for "still doesn't work."

- Anonymous, journaldev.com

## Case Study #1: Complacency Fuels A Disaster

### Background

Late summer in western Ohio is beautiful: brilliant deep orange sunsets over a flat simmering horizon of farmhouses and silos and rich corn fields.

But in 2007, away from the sanguine setting, inside a large, dark refrigerated warehouse, an exhausted project team was struggling to convert the operation to new inventory management software. An operational disaster was unfolding. Manufacturing lines in the plant, normally producing and shipping 50 truckloads a day, had come to a halt, blocked by inventory waiting to be shipped that choked every available space in the warehouse.

Things turned into a heart-stopping disaster as inventory piled up and eventually shut down production lines because there was no place to put product. The outbound deliveries to distribution centers also slowed, resulting in trucks delivering one to two days late. Eventually the customer felt the disruption as product failed to make it to the distribution centers in time to fill waiting orders.

Displaying my keen sense for career-enhancing moves, I got into a heated argument with the VP of Operations. We disagreed as to the true cause of the disruption. The stress on our teams was enormous. Although production lines were able to start back up after a day, we still had an inventory-choked warehouse in which movement with a forklift was difficult, and every task proceeded at about half the normal speed. Day after day we struggled with how to get enough breathing room to clear away the overstock in the building.

A general sense of profanity-laced panic was setting in. How had it come to this?

It took a solid month to dig out. But not before making customers angry, incurring hundreds of thousands of dollars in obsolete inventory losses, and completely burning out the 30 or so people who worked day and night on the problems. I guess compared to other, more epic disasters, I should have been elated.

No one thought that this launch would be any different than the previous two launches we had done in two of the company's other plants that year. Those other two implementations had gone well, with no disruption to the business and with general stability after about two weeks.

In IT project terms, having to shut down a large plant because of troubles cutting over to a new system is considered one big fat failure. Senior management was livid.

## Key Points

**The team was complacent about the software.** We had "gone live" in two other plants that year with the new software. What could be different about this plant that would create problems? The Ohio plant, although bigger than the other two plants, made the same product, and shipped it in the same way, as the other two plants, the software was configured in the same way, and we had thoroughly tested the application as if production and shipping were coming out of the Ohio plant.

But the Ohio plant was different in at least one important respect: its production rate in pallets per hour was much higher, and that meant forklift operators would pick up in a random fashion any pallet they could at the wrapping station in order to keep up. They didn't necessarily pick up and scan in a first-wrapped-first-pickup fashion, and that went against the logic of the software (this was news to us).

For some reason, in the other two plants, the forklift operators would always pick up and scan the pallet in the queue that had *first come off the wrapping station.* So we hadn't encountered any problems.

We discovered that the system operated according to a kind of First-in, First-out rule, which we never expected to be applied to this particular pallet movement. Surprise! The operator could still move the pallet but had to access a different transaction on a different screen on the truck-mounted radio-frequency (RF) scanning unit. This slowed down the entire flow of product.

So we had let our complacency about the system override any thoughts that the Ohio plant might have different operating characteristics that the software would react differently to. We were caught by surprise on an issue that significantly and negatively affected our launch.

The impact from the additional steps in the new system was completely underestimated. Coupled with operators fumbling with handheld scanners and RF terminals, our throughput capability fell drastically. In reality the organization was unprepared for the new system. As a result everything slowed down and soon the normal flow path through the warehouse was clogged with inventory.

What did we learn, and what would we do differently? "Throughput" is now one of my favorite words. If you can't demonstrate that throughput can be equaled or exceeded post-new system implementation, you have a major risk that must be fixed. We also learned that training classes are insufficient to prepare a workforce for a new system. Instead, a *change management* process is required, in which you examine each change brought on by the new system, how it affects a person's daily work, and what the team is going to do to ensure those changes are brought on without disruption to the business.

**The team was complacent about the readiness of the organization**. The Ohio plant was large, yet the staffing in the warehouse was rather thin. We had held numerous training sessions for weeks, and our point person for making sure the work force was ready assured us that training had been sufficient. But the truth was something different.

The Ohio plant produced 24 hours a day, seven days a week. Each pallet that was produced, at a rate of one every 45 seconds, required 3 to 4 bar code scans, each representing a transaction in the system requiring a user to key in data, before it was shipped. Prior to the new system, no bar code scans were required, and no data entry was required until the truck was ready to leave. This allowed a relatively small group of operators to ship 50 truckloads in a 24-hour period, because they could freely move pallets around without scanning and recording movement.

The team didn't realize how the extra steps from scanning and data entry would slow down work. The warehouse team had never used bar code scanners before. The steps were simple, but the process involved pressing buttons on a very small keyboard, scanning a bar code label, and then pressing other buttons. The lighting in the warehouse wasn't great, and wearing gloves while operating the scanners felt clumsy. All of this combined to slow down the whole operation.

In retrospect, we should have sent a few warehouse operators to one of the two other plants we had already successfully converted to the new system. The operators would have had a chance to work in a live environment, actually performing the work instead of just learning about it in a classroom. These operators could have also served as trainers in the Ohio plant.

## Case Study #2: Nike's Debacle With Supply Chain Software

### Background

In February 2001 Nike, Inc. announced that it would miss sales and profit targets for the quarter due to problems with supply chain software it had begun to implement the previous year. The company said that it had experienced unforeseen complications with the demand and supply planning software that would result in $100 million in lost sales.

Nike was trying to put in a system that would cut its response time to changing sales demand. These types of systems rely on algorithms and models that use historical sales data combined with human input to generate a sales forecast, which is then converted to a manufacturing plan and orders for raw materials from suppliers. It's not easy to set up and successfully run these applications to produce optimal results. The process demands a lot of trial and error, testing, and running in parallel with the old system to shake out bugs.

As reported by CNET News' Melanie Austria Farmer and Erich Leuning, SAP spokesman Bill Wohl, reflecting on Nike's dilemma, said at the time, "What we know about a software implementation project is that it's just not about turning on the software. These projects often involve really wrenching changes in a company's business process...It involves changes in the way employees work, and anytime you make changes in the way employees are used to working, it can get difficult."

Nike is in the apparel business, where styles come and go, and where advertising and promotional programs can spike demand, requiring the supply chain to react just in time, delivering to the market just the right amount of each style. An oversupply of shoes or other apparel will lead to discounting and reduced profits, and an undersupply will lead to lost sales. Nike ran into both of these scenarios, and its profit dropped while sales declined, resulting in the $100 million unfavorable financial impact to the company.

Inside the logic of the software Nike chose, parameters and settings must be optimally set for the most efficient quantities to be produced and distributed to the market. It's very easy to get it wrong, and companies launching this type of application usually run a pilot for several months before they are satisfied with the recommended production and distribution plans generated by the software.

Much has been written about Nike's experience, and much of it is valuable for any enterprise thinking about a similar project. Keep in mind, though, that this was a public spat, and both the software firm and Nike told their own version of the story for the public record. That means we don't have all the facts. Nonetheless, I think there are valuable lessons in the Nike story, and at the risk of not getting all the facts right, I present my conclusions more to help you learn and succeed than to cast blame on any of the Nike project participants.

### Key Points

Here is what I think were the main issues in the Nike project:

**Complexity of the application without commensurate resources applied to making it work**. Christopher Koch, writing in CIO Magazine at the time, said "If there was a strategic failure in Nike's supply chain project, it was that Nike had bought in to software designed to crystal ball demand. Throwing a bunch of historical sales numbers into a program and waiting for a magic number to emerge from the algorithm – the basic concept behind demand-planning software – doesn't work well anywhere, and in this case didn't even support Nike's business model. Nike depends upon tightly controlling the athletic footwear supply chain and getting retailers to commit to orders far in advance. There's not much room for a crystal ball in that scenario."

I don't fully agree with this assessment; I think demand fore-casting systems are critical to modern businesses, and if config-ured and used correctly, bring many benefits. Other reports said Nike didn't use the software firm's methodology, and if true, this would greatly contribute to its troubles. I have implemented these systems and they require precise attention to dozens of settings and flags, pristinely accurate data, and the flawless sequential over-night execution of sometimes 30 or more heuristic calculations in order to produce a demand forecast and a recommended produc-tion and raw material supply plan.

It's also critical with these types of applications to have the right subject matter experts and the best system users in your company on the team dedicated to making the system work the right way for your business. This is where, if published reports are true, I believe Nike may have failed. It is possible Nike simply needed more in-house, user-driven expertise, and more time to master the intricacies of the demand planning application.

In 2003 I ran an ERP project that included an overhaul of sup-ply chain systems. The suite included demand and supply plan-ning solution software, which we would use to forecast demand, generate a production and raw materials supply plan, and deter-mine the plan for supplying product from plants to distribution centers. Unfortunately the best system users declined to be part of the team due to heavy travel requirements, and we had multiple problems getting the parameters right. The supply chain suffered after launch as incorrect production and distribution plans dis-rupted the business for several months.

Combining a maintenance-heavy, complex application with an organization unwilling or unable to meet the challenge is one way to find the money pit.

**A 'big bang' approach to the launch without sufficient testing.** Despite prevailing wisdom and suggestions by veterans that Nike phase in the new application, Nike chose to implement it all at once. This immediately put at risk a large portion of the Nike business. A phased approach would have limited the potential damage if things went wrong.

A case study of the project published by Pearson Education discusses this point: "Jennifer Tejada, i2's vice president of marketing, said her company always urges its customers to deploy the system in stages, but Nike went live to thousands of suppliers and distributors simultaneously"

The study also quotes Lee Geishecker, an analyst at Gartner, Inc., who said "Nike went live a little more than a year after launching the project, yet this large a project customarily takes two years, and the system is often deployed in stages."

Brent Thrill, an analyst at Credit Suisse First Boston, sent a note to his clients saying that because of the complexities he would not have been surprised if to test the system Nike had run it for three years while keeping the older system running. According to Larry Lapide, a research analyst at AMR and a supply chain expert, "whenever you put software in, you don't go big-bang and you don't go into production right away. Usually you get these bugs worked out . . . before it goes live across the whole business."

I can understand that Nike would want to convert a large portion of its business and supplier base at the same time. It reduces the length of the implementation and therefore the cost of maintaining consultants and support staff, and it eliminates the need for temporary interfaces to existing systems.

But a smart move might have been to launch and stabilize the demand planning portion of the software first. It's easy for me to second guess, but Nike could have taken the forecast generated by the new system and entered it manually into their existing, or 'legacy' systems. After all, if the forecast is wrong, then everything downstream – the production, raw material, and distribution plan – are also wrong. I did this on two projects, and it significantly reduced risk. On both projects we launched the demand planning (DP) application and ran it in parallel with our legacy system until we were satisfied with the results, then we disengaged the legacy DP application and began manually keying the new system's DP forecast into our legacy production, raw material, and distribution planning software.

## Case Study #3: Denver Airport Baggage System

### Background

"Denver Airport" has become synonymous with epic technology failure for those who remember the colossal breakdown of that airport's ambitious new automated baggage handling system in 1995. Just two numbers explain the magnitude of the failure: 16 months – the delay in opening Denver's new airport – and $560 million – the extra cost of building the new airport – both a direct result of the baggage system debacle.

Denver Airport has huge lessons for us that can be applied to any software endeavor.

In 1993 a brilliant plan was hatched to fully automate baggage handling at Denver's new state of the art airport. The key to the automated system was software that would control the movement of thousands of carts moving along miles of track throughout the airport's three main terminals. The software would direct the movement of the carts to collect or deposit bags at precisely the right time. It was to be a highly orchestrated activity that depended on software that would continuously process complex algorithms. After more than a decade of trying to make the system work, the airport went back to the traditional manual bar code tug and trolley system.

**Key Points**

**Hubris can lead to very bad decisions.** The airport's new system was to be state of the art, the most automated passenger baggage system in the world. Through 26 miles of underground track, bags would move from plane to carousel or gate to plane without human handling. Tours were given to the public to show how advanced the new system would be.

BAE Systems, the company hired to design and build the baggage system, had built several traditional systems at airports throughout the world, but none were as advanced as the Denver project's design. A sign of early trouble came when the airport bid the design out for construction, and several companies either declined to make a bid or responded that construction of the complicated system within the airport's stated timeline was impossible. The Denver team was proud of its futuristic design and even these clear signals of danger ahead did not dissuade them from their plan.

**Complexity greatly increases risk of failure.** All along the miles of track of the new system, thousands of small carts would deposit or pick up baggage at precise points in the network, ostensibly at just the right time. This required complex algorithms that had to account for travel distance, expected flight arrivals and departures, sorting rules and routings, and canceled flights. Scanning devices positioned at just the right locations would read bar code labels applied to bags and route them to the appropriate conveyor. This type of technology, while standard today, still isn't perfect; you can imagine its relative immaturity in the 1990s.

**A 'Big Bang' launch of a new system adds to likelihood of failure.** Denver's airport authority had a great opportunity to start small and prove out its advanced system design. The team could have sliced the project into digestible parts in two ways: An end-to-end prototype system on a small portion of the airport's baggage traffic, or a fully functional piece of the envisioned architecture, such as the bar code label and scanning functionality. Small pieces are easier to focus resources on. Best of all, had the Denver team phased in the system gradually and still faced failures, it wouldn't have shut down the entire airport. Being able to continue business as normal if the new system fails is an essential but sometimes forgotten aspect of all software projects. Building in pieces or parts allows new learning to be incorporated into the design. Instead, the team gambled on launching the whole system at once.

**No backup plan = nasty outcome.** Once the Denver team realized that it may take awhile to make the new system work, they rushed to put in place a more traditional trolley system using baggage handlers. But this alone took many months and an extra $70 million before it was completed. In the end, the original advanced-design system was only ever used for outbound baggage at one terminal. Large parts of the airport's new system simply never worked. But a failure or cost overrun on the original ambitious project is one thing. Because there was no backup, the project was squarely in the critical path of the new airport's opening. If you're going to fall out of the boat, don't drag everyone with you.

OK, you say, your organization isn't launching anything nearly as ambitious as the Denver Airport baggage system. And of course you would never make the dumb mistakes they made, right? Yes, you would, because every human being makes mistakes; the Denver team just put a lot more at stake. It's not so much that their software design failed, it's that they placed a huge bet on their system working – the opening of a new $3 billion airport.

## Case Study #4: Did We Get Anything Out Of This?

### Background

Company X processes 50,000 customer orders every year. To do this, they employ 30 customer service representatives in a centralized department. In January of 2011 the company decided to invest in software modifications that it said would cut thousands of man-hours of administrative work.

The order processing 'process' had multiple steps, and involved having to manually check each customer order even though nearly all orders were sent electronically directly to the company's ERP system. Order changes were also cumbersome, and the company had no systemic way of 'allocating' available inventory to specific customers in cases where product was in short supply.

A lot of work went into calculating the projected savings in reduced administration. Automation of transactions would significantly reduce the time required for daily activities in the system. For example, a 'no touch' order process was designed so that the system would perform the checks that human beings would have done.

The IT experts in Company X and its contract developers were consulted to determine the cost of the software modifications, using a list of 12 modifications the customer service team had drafted. Nearly all of the cost would be for the ERP software developers' time and for a consultant with ERP design experience who could work directly with the customer service team to interpret their needs and translate them into specifications the developers would need to complete the programming.

The estimated cost was $400,000, not a huge sum for Company X, and relatively small by software project standards. The team estimated it would take six to eight months to complete all 12 enhancements, but they would be phased in as soon as they were completed.

This expense was to be considered a capital investment, since software costs, for tax purposes in the United States, are normally depreciated over a three-year life. So the team prepared a capital expense request with the requisite project benefits and costs. The team proposed that while actual labor costs would not decrease, the huge reduction in administrative work would allow the customer service team to pursue more "value-added" work, such as helping the customer determine a more efficient order size and order frequency. Also, Company X had an aggressive growth plan – sales were growing at about 20% per year – and the team argued that the reduction in administrative work made possible by the software modifications would allow the company to grow without adding headcount.

Company X approved the $400,000 capital investment. It was April 2011 before the right consultant could be available to launch the effort, and he split his time between New York, where Company X was headquartered, and Texas, where the customer service team was based.

The first enhancements were delivered in the fall of 2011, and they worked as designed by the team. But they represented only about a third of the total modifications needed. It took until late summer of 2012 to complete the biggest modification: systematic allocation of available inventory to specific customers selected by the customer service team. Even then, while this enhancement was complete, it still had to be thoroughly tested, and this took until the end of 2012.

The costs for the project rose to about $625,000, due to unforeseen complexity in the inventory allocation enhancement and the extra months of the consultant's time and travel expenses due to the project taking longer than expected.

The customer service team is using the modified software and some team members report solid efficiency improvements in their work. The team is the same size as before the project began. None of the customer orders can be processed as 'no touch' and many require several touches before being released into the ERP system.

While the system changes produced some efficiency, it was not substantial enough to make a clear difference in generating more valuable time for the team.

### Key Points

**From the outset, the project's ROI was non-specific and "soft,"** meaning no hard dollars were identified, just reduced hours of administrative work. The company didn't get specific commitments for cost reduction, sales increases, or other tangible benefits, just hours saved that could be invested in more valuable work.

As a result, the improvements from the new software are hard to see and measure. Some who work with the new changes say they are time-savers, but what is done with the time saved has only been anecdotally communicated, not measured and verifiable. The customer service team is, however, able to process about 10% more customer orders per person than it could prior to the software enhancements.

Processes didn't change to line up with the new capabilities, primarily due to a lack of commitment to change. Users were involved in identifying inefficiencies, in finalizing the specifications for the changes, and in testing the developments. But there was not a commensurate commitment to change the way the team processed orders. As a result, things became more efficient in a narrow sense, whereas a pre-project commitment to new and more efficient processes could have resulted in realization of greater benefits.

**Poor initial estimates of cost were** caused by a lack of understanding of the level of complexity the changes would require. Not enough time was spent thinking through the changes that would have to be made to the ERP system. The developers working on the project did not have experience with some of the desired changes, so there was a level of creativity and inventiveness that had to take place. This made it difficult to nail down accurate cost estimates.

**A lack of cost control,** both contractually with the developers and management-wise with the project team, allowed costs to creep upward without much resistance. The team did not regularly review costs incurred to date and estimated costs to completion. Often the increases were not disclosed until the developers submitted invoices.

## Case Study #5: The 2010 Census

### Background

The 2010 census was to be the most automated census ever. Half a million census takers with handheld devices would capture, in the "field," household data that had not been sent in via census forms. Harris Corp. was given a $600 million contract to supply the handheld devices and manage the data.

But according to published news accounts, the Census Bureau requested, after the project began, more than 400 changes to the requirements that had originally been submitted to Harris. In trying to accommodate these requests, Harris naturally encountered more expenses to redesign or re-program the handheld units or to redesign the data management system that would collect and organize the accumulated data.

The handheld units themselves were difficult to operate for some of the temporary workers who tested them, and they couldn't successfully transmit large amounts of data. A help desk for field workers using the devices was included in the original contract at a cost of $36 million, but revised to $217 million.

In the spring of 2008, the Census Bureau was faced with a decision whether to continue with the automation plan, because the handheld units had not yet been completely tested and needed further development, in part because of the additional post-contract requirements. The Bureau needed enough time to hire and train about 600,000 temporary workers if the original Field Data Collection Automation (FDCA) plan had to be revised or scrapped.

In the end, the 2010 Census may not have been the most automated census ever, but it was the most expensive. The contract with Harris was revised to $1.3 billion, and other expenses were incurred for equipment and other areas that were not anticipated and therefore not estimated. Not all of the overruns were systems-related.

### Key Points

**Constantly changing requirements increased delays and costs**. As we know from understanding the nature of software, a system is unable to simply change its code and accommodate additional requirements on the fly. Why no one put a stop to the additional requirements heaped on to the project is a mystery, but it's pretty much standard procedure to freeze the requirements at some point in the project. It's like asking a home-builder to add another bathroom on the second floor when the home is halfway to completion. It can be done, maybe, but will make the house cost more and take longer to complete. In extreme cases – like the new custom-built Medicaid claims processing system for the State of North Carolina – the project may never end.

**Undue confidence in the user's ability to learn how to operate the handheld devices led to surprise additional costs**. The project didn't plan on people having so much difficulty with the handheld data collectors. But people's innate abilities, especially in the area of new technology, vary greatly. Nearly every project I've been involved in experienced difficulty because of a certain percentage of users not being able to catch on to the new system. This means more mistakes are made with the new system, more support is needed, and in some cases people who were competent at their jobs with the old system simply cannot perform at a satisfactory level with the new one.

**The project ended up being a money pit**. The Census Bureau had to revert to pencil and paper when the handheld devices couldn't be used – which it said would add $3 billion to the cost of the census. If $3 billion is what the Bureau would have saved with automation, then it was probably worth it to invest the originally estimated amount of $600 million, and even the revised estimate of $1.2 billion. Instead, the government paid the full $1.2 billion *and* had to use pencil and paper. Net result: a waste of money.

Just freezing the requirements alone, at some point in the project, could have completely changed the outcome. Intentions were apparently good – saving labor cost through automation – and I expect there were presentations made to different levels of management in order to gain approval. *A well-intentioned project developed by smart people becomes a vast hole sucking time and money into the abyss.*

## Case Study #6: Customization Surprises

### Background

It was late fall, 2005, at the headquarters of a well-known company we'll call Company W. The selection of the (fictitious) Really Great Warehouse Management System (RGWMS) for the company's three manufacturing plants was a sound one, everyone felt, in part because it would cost about $1.3 million less than the industry-leading packaged applications. The company planned for implementation to begin after the first of the year. This would allow time for formal approval of the project's $1.8 million budget.

I had come on to the project in February 2006. Much of the spring and summer of that year was spent running the software through hundreds of business scenarios to identify "gaps" between how the warehouse would process product movements in the warehouse, such as pallet put-away, picking, staging, and shipping, and what the software could perform in its current "out-of-the-box" state.

The project team began logging a significant number of gaps, which is normal for a packaged software application. But one area of missing functionality was particularly maddening: the limited ability of the software to process radio frequency scans of the barcoded label placed on each pallet of product.

For each pallet, we had designed four to six scans as the product was received from the manufacturing side of the building. Each scan would tell RGWMS where the product was, where it moved from, and where it was supposed to go. This particular version of RGWMS, however, did not process scans the way we had expected a normal warehouse management application would. When one of the consultants explained how this version of the software processed scans, I had three words: You. Are. Joking.

It seems that for every movement of a pallet inside the warehouse, RGWMS expected that a Transfer Order, or TO, would be created, by, say, a warehouse supervisor using the system. Then, when a forklift operator scanned a pallet to move it from point A to point B, he was merely "confirming" the TO that had already been set up in the system by the supervisor. This meant that a facility producing 2,000 pallets per day, with each pallet being scanned four to six times before being shipped, would require someone to enter 8,000 to 12,000 TOs into the system each day, one at a time.

Maybe there was a warehouse out there, somewhere on the planet, that could use this type of scanning process, but for our project it was completely unacceptable (and maddening!). So we invested hundreds of thousands of dollars to automate the TO creation process, mainly by creating the TO "on the fly" whenever an operator scanned a pallet. When a pallet was scanned, a simultaneous transaction in the background would create the requisite TO.

The result: about four extra months to complete the changes to the software, pushing the three launch dates out and adding unplanned expenses for programming and extended engagements for the project's consultants. The total cost came to $3.1 million, nearly exactly what it would have cost to implement either of the two other industry-leading applications the company had considered.

### Key Points

**Development costs were under-estimated, due to insufficient exploration of the particular version of RGWMS the company had chosen.** A conference room pilot (CRP) would have been a good way to more thoroughly identify the version's functional gaps. In a CRP, the software is loaded on to a test server, configured the way the enterprise will use it, and loaded with some of the company's real data. Once this is complete, users spend a few days performing the transactions they would use in the test system, and along the way gaps – missing functionality – are noted. Usually the vendor's technical team is also present so as to explain how the software works in each step of the process.

**Software does not respect rules of common sense**; it only knows what the programmer wrote in its code. Just when you think you understand how an application works, it throws a monkey wrench into your carefully ordered world, nixing your plans and vacuuming up money you didn't budget. In this case company W thought it knew enough about RGWMS to estimate implementation costs.

There is a tendency in cases like this to get the project moving, and so assumptions are made – almost subconsciously – about what the software will provide and how it will perform. These assumptions almost always lean to the optimistic side; that everything will somehow work out fine.

# 3
# WINNING CASE STUDIES

"How you play the game is for college ball. When you're playing for money, winning is the only thing that matters."

- Leo Durocher

## Case Study #7: Conversion To SAP In 90 days

### Background

Company Z - U.S. (ZUS) is a $1.3 billion maker and distributor of over-the-counter pharmaceuticals based on the East Coast. ZUS is a subsidiary of Company Z – Global (ZGLO) based in Western Europe. In 2011 ZUS acquired northern California-based Company M, a $100 million pharmaceutical company with a plant and two warehouses in northern California and another in Spokane, as well as a cross-docking facility in Texas. Company M had been growing in recent years at more than 35% a year and ZUS, by acquiring Company M, sought to gain market share, product diversification and expanded distribution.

Company M grew out of a family business and relative to ZUS, its systems, controls and processes were rudimentary. Company M had added software applications where needed, yet still performed a lot of work manually. It had packaged software applications for financials, raw materials, production planning, warehouse management and finished goods inventory.

ZUS, however, had been managing its business for over 10 years with SAP. It used SAP for order processing and shipping, demand planning, product deployment and production planning, raw materials management, financials, master data, and finished goods inventory management. It used a configuration of SAP largely defined by its parent company ZGLO, and reported financial results to ZGLO in accordance with standardized SAP cost centers, general ledger and chart of accounts.

As ZUS began to integrate Company M into its management processes, it realized that financial reporting from Company M was limited and manual. Having managed its operations using revenue and expense measures defined in a certain way in SAP, ZUS could not accurately interpret Company M's financials in the same way and could not easily compare Company M's business performance with that of ZUS. ZUS concluded that it had to institute within Company M the same financial definitions, controls and methods of reporting that it used already, and that meant converting Company M from its current systems to SAP.

While the primary ROI of the project – converting Company M to SAP would cost about $1 million – was the ability to financially manage the $100 million company, ZUS went further and defined specific "value realization" targets that would result from the conversion. Included in this were hard savings from staff reduction because SAP would automate certain manual processes.

A strong mandate by ZUS's CEO mobilized a small team in the fall of 2011 that would determine how to quickly bring Company M onto an SAP platform. One of the first and most important decisions was project scope – what absolutely must be converted to SAP and what can wait? The team chose a scope sufficient to manage all core transactions in SAP while leaving to legacy systems or manual work those things that weren't essential to "managing" the business in SAP.

Nonetheless, the team had to convert to SAP hundreds of products, customers, raw materials and vendors, and the purchasing, material requirements planning, order processing, manufacturing, shipping, invoicing, cash management, payroll and financial functions. The team decided to use as much of ZUS's SAP footprint as possible, and to try matching business processes of Company M with those of ZUS.

The team had a project manager with 10 years of SAP experience and six to 10 full-time or nearly full- time business process and IT experts. Getting adequate time from Company M's business process owners was extremely difficult because the organization was so thin, with Company M being a much smaller company than ZUS.

The cutover was completed on March 4, 2012, exactly on time and approximately $300,000 under budget. It had taken 90 business days from start to finish.

### Key Points

**A limited scope made a quick implementation easier.** The team matched the scope to the time allotted for the project, without losing the integrity of having all core functions running on SAP. For example, not all customers were converted to EDI for order receipt, and not all of Company M's third-party distribution warehouses were integrated with advanced ship notifications (ASNs). These weren't needed to get a complete picture of the business in SAP.

**Lengthy configuration debates were kept to a minimum by using ZUS's SAP footprint.** Questions like "what storage locations should be defined in the manufacturing plant?" and "how should we define product, brand, packaging format, and UPC in the system?" were largely already answered by using definitions and parameters (collectively called set-up) that ZUS already used to run its business on SAP. For example, at ZUS, the product "hierarchy" went like this: "brand" is at the highest level and can include many "pack groups," and pack groups can include many "formats" or packaging configurations, and finally, the product itself, the UPC (or SKU) fell underneath the pack group. This sounds like a tedious point, but it's these kinds of decisions which can take a long time for enterprises to make. The longer a project takes, the more expensive it will be.

**Overall, success was made possible by a small, strong, experienced and focused team with a CEO mandate.** Most of the team members had SAP experience. Some had led previous implementations and many had deep functional knowledge of a particular area or areas within SAP. The project manager on several occasions threatened to move out the launch date – and invoke the wrath of ZUS senior management – if Company M's business experts did not adequately engage in the project.

Nearly all of the project work was done on-site at Company M California headquarters. This allowed easy access to Company M subject matter and process experts, and made it possible for them to experience the software during testing. More importantly, the on-site work – usually in one large room – made for instantaneous communication and sometimes instantaneous problem-solving. *There is nothing more powerful in an enterprise software project than knowledgeable humans being in the same room all focused on the task at hand.*

ZUS drew on its experience with SAP to provide go-live support. SAP users at ZUS were brought in for the first three weeks of the cutover to help Company M learn the application and at the same time keep the business going. This included a "buddy" system, or pairing of one ZUS user with one or more Company M users, especially for critical functions like production, order processing and shipping.

## Case Study #8: A Win On Dashboard Software

### Background

Company Y measures its logistics performance via several key performance indicators, or KPIs. These KPIs measure truckload utilization, on-time delivery, customer order fill rate (quantity shipped divided by quantity ordered) and several other metrics, by manufacturing plant, distribution center, brand, customer, sales region, channel of business and total company.

For many years, the raw data for these KPI reports came from several different software systems. People would extract data from these systems into Excel spreadsheets, where calculations were done, merged with other data, charted to show trends, and then emailed to managers and other users of the reports. Sound familiar? The process was time-consuming, manual and fraught with errors and redundancy. People in different parts of the company, for example, were all calculating and publishing customer order fill rate every day, week and month, duplicating efforts. Also, not everyone came up with the same calculated metric, although they should have, since everyone used the same underlying data.

The situation seemed to call for a rationalization of the process and some type of IT project, right? Maybe. In the summer of 2009 Company Y first turned to the same idea many companies think of – a centralized data warehouse. A data warehouse could capture all of the transaction data from the system of record – In this case SAP – and provide the same data to everyone in the company via a variety of queries and reports.

To build a data warehouse, and then enable reports and queries of the data based on user preferences is no small effort. Just to determine the design of the warehouse is a huge undertaking. For example, common data definitions have to be established (what is a "customer" – the warehouse that orders the product, the entity that pays the invoice or the buying entity that places the order?). Also, different parts of the company must agree on relationships between the data. For example, each customer must be "attached" to a sales region, a sales manager and a price list.

Company Y could have pursued this data warehouse route, and probably with a lot of time and money would have been successful. But Company Y didn't have that kind of time or money, so it sought a different solution.

By 2009 dashboard products had been on the market several years. These types of applications can take data from disparate systems, join that data according to user definitions, and display it in numeric, table or graphical format according to user-selected views. The result is something like an Excel spreadsheet with pivot tables, except that all the tables are already built and the data links are set up to run in the background. Company Y wanted a "dashboard" that would display the important KPIs every day. The dashboard would be the single version of the truth company-wide.

Company Y looked at solutions offered by the top-tier business intelligence (BI) and dashboard software providers, such as SAS, Microsoft, Information Builders and Cognos. All of these companies provided sophisticated BI environments that were overkill for what Company Y needed, so Company Y considered applications from smaller and newer firms that offered a simpler and less complex approach.

In a decision I would characterize as a smart IT investment, Company Y chose software from Qliktech Inc., a company whose marketing message is: "Traditional BI solutions have become bloated, complex software stacks, leaving users confused and frustrated. For 18 years, QlikTech has focused on simplifying decision making for business users across organizations."

Qliktech's flagship product, Qlikview, is basically a front-end aggregator of disparate data sources with graphical display capabilities such as charts and color-coded alerts. It is simple and intuitive for the user, and users can be trained to develop new dashboards on their own. Company Y purchased a server license and negotiated a one-time per user license. The combined cost was orders of magnitude less than building a data warehouse.

Company Y didn't have to build a formal data warehouse because Qlikview could access data from Excel files, tables in a database or Microsoft Access. Company Y had already been capturing data from its core SAP transaction system for basic sales and cost reporting. The company stored this data in existing tables in an Oracle database. It was relatively simple to refresh the data in Qlikview each day from these tables.

Company Y continued to expand its use of Qlikview. Three years later, the company is using Qlikview for sales reporting, promotional funds spending, nearly all of its supply chain KPIs, transportation cost monitoring, some demand forecasting metrics, finished product inventory flow monitoring, and raw materials inventory management. More than 200 people use one or more of some 60 active dashboards.

Users give high marks to the application because all the data sourcing and calculations are done automatically every day, and they can slice the information in several different ways without running pivot tables or macros in Excel. Management is happy with the relatively low cost and high flexibility of the solution. Some users have learned to program their own dashboards; one user has developed a dashboard that runs macros automatically to create a day's worth of purchase orders in SAP.

### Key Points

Company Y avoided a huge expense and many months or even years of development work by choosing a relatively simple application. The company was smart enough to learn first what solutions the market offered, and knew that costs should be kept to a minimum because the returns were "soft" – administrative time saved by existing staff, not a true reduction in head count or labor cost.

Over the past three years Company Y has spent a fraction of the cost of a data warehouse yet equipped users with a robust set of dashboards that are used regularly to manage the business. The

Qlikview dashboards at Company Y are in fact the single source of some important metrics. Moreover, the dashboards are centrally located so users who want to develop a new set of analytics can first peruse the 60 active dashboards to see if any of them fit their needs.

Company Y also made use of its existing data architecture, which, although not in any sense a data warehouse, had taken years to build. The data capture – the hardest part about implementing applications for analytics – was already done. The company didn't have to re-invent its methods of retrieving and storing transactional data.

The project was also a success because it had a limited scope that satisfied the main business need, and was strategically wise because the low investment didn't lock the company into a specific solution whose payback might have been years into the future.

## Case Study #9: SaaS Success Story #1

### Background

The cold spray flew up and atomized in the chilly air from the bow of the 25-foot boat as it rode the swell and chop of Lake Michigan on a clear and bracing fall evening. Our boat bounced over whitecaps toward the sunset mural of orange and purple sky on the western horizon. Not bad for a business trip. At the helm was a member of the executive team of a small software company, Company A.

Company A sold a transportation management system (TMS) that determines optimal routing, carrier selection and truck utilization. The unique aspect of Company A's TMS was that firms did not have to purchase and install the software. Instead, customers accessed the system from a web portal, and paid for the system on a monthly fee basis.

I was with Company G, and we had spent half the day and an evening with the company's entrepreneurial team, having just contracted with the company to use their application.

The decision to select Company A was the result of a typical cost-benefit analysis among several alternatives. The main alternative was to stay on the TMS Company G had at the time, a packaged application the company had acquired several years ago. But a decision had to be made because the support contract on the existing TMS application was due to expire, and the only way to extend the contract was to upgrade to the next version. The version upgrade would require a capital investment plus the normal annual maintenance fee of 20-25% of the initial license cost. Included in the capital expense would be the license, implementation consulting fees, and an operating system (OS) upgrade on Company G's server. Company G would continue to host the application in its data center and staff people who could address server and software issues.

The costs for Company A's Software-as-a-Service TMS were a small implementation fee covering the cost of one of their technicians for several weeks, and a flat fee per month, plus costs of building interfaces to SAP. So in simple terms the options looked like this: Option 1 was a significant capital investment plus annual maintenance fees plus server maintenance and application support. Option 2 was a small upfront fee plus a monthly fee.

It looked like Company G could save on yearly costs by staying with its current TMS and upgrading, except that it would take 10 years for the annual savings to pay back the initial capital investment. And during that time, Company G was likely to invest in at least one if not two more upgrades, which would only extend the payback.

So Company G signed a contract with Company A. The implementation was completed within three months, with no disruption to daily shipping activity. The company quickly expanded the number of users who could have access to the TMS, because the license was an enterprise license with no limits on number of users, and because the shipment data in Company A's TMS was up-to-the minute and valuable to a large number of people, including Company G's third-party distribution centers, manufacturing plants and customer service team.

**Key Points**

**The company was aware of new applications available on the market**. Company G's transportation team, led by its manager, had already investigated different application options because some members of the team participated regularly in industry share groups, where participants could share best practices and experiences.

The Software-as-a-Service option was a clear winner on paper – and it turns out, in actual results – in terms of costs and functionality. The traditional payback assessment had revealed a lifecycle cost of the SaaS option that was less than the typical client-server on-premises application. Any benefits from improved functions and features – and there were many – came on top of the overall favorable life-cycle costs.

**The SaaS application could be used by anyone in the enterprise** – not just transportation planners. An entire network of people working within the supply chain suddenly had access to important information that affected their daily objectives. Distribution centers, for example, could see the status of trucks heading inbound to their facilities. Company G also had a pressing need for additional reports, a way to electronically capture shipping discrepancies, and the ability to create upload files from data in the TMS application. All of these were provided by Company A at relatively low cost.

**The software vendor's revenue model was in harmony with the company's financial goals**. The savings in ongoing support and development costs have been tremendous. Company G has been able to avoid capital expenditures. The expense of the software is directly related to the function that uses it. Upgrades to the TMS are all managed behind the scenes by Company A.

This case study is an illustration of how good a SaaS application can be. And if there is one good SaaS TMS, there must be other good SaaS applications in other functional areas and industries, which is a way of telling you that the game has changed and you don't have to live with traditional client-server applications that you own and maintain. More on this later in the book.

## Case Study #10: SaaS Success Story #2

### Background

When retail stores started using bar code scanning devices at checkout counters, companies emerged to collect and sell that point-of-sale (POS) data to a retail chain's suppliers and other manufacturers of products sold to retail outlets. Nielsen has built a worldwide business supplying POS data and analytics.

In the 21st century, retailers themselves are collecting and analyzing POS data with their own systems; some are offering to share it with or sell it to companies whose products they sell in their stores. Many manufacturers want this data and believe significant insights can be gained, such as how often and when their products are out of stock in stores and how special pricing or incentives affect sales.

But processing that data is easier said than done. Let's say 3M wants to analyze the sale of its tape products at Home Depot stores in the U.S. Every single sale of a roll of tape would be captured by Home Depot's POS system, recording the price, the SKU (stock keeping unit) – whether it is 1-inch masking tape, 2-inch masking tape, a 3-roll pack, blue painter's tape, electrical tape, packing, tape, or one of the other many variations of tape sold by 3M, the date, the store number and address, and other pertinent information, for each of Home Depot's 1,974 U.S. stores.

This represents a flood of raw data – like drinking from a fire hose. Few companies have tools with which to analyze this enormous volume of data. Some are building their own data warehouse; others are considering packaged business intelligence applications.

In late 2010 a well-known brand name company – Company H – was weighing its options for analyzing its products' POS data from one of its largest customers, Really Big Retailer (RBR). The data would be in raw form; about 9 million lines of data per week. For several months the company investigated storing the data in a data mart and using query tools to analyze it. Some of Company H's staff tried using Excel spreadsheets on smaller samples of the data. At that same time, Company H's sales team investigated outside software options.

Company I, a firm that specializes in analyzing retail POS data, had reached an agreement with RBR to collect, organize and sell RBR's store sales data to RBR's suppliers. Company H had been in discussions with RBR on how best to gain more insight to shelf performance and shelf conditions.

So Company H turned away from building something on its own and instead negotiated an agreement to purchase Company's I's POS analytics software on a pay-as-you-go basis. Company I's model is to sell proprietary analytics, via a web portal, of the POS data it obtains. It sells reports, analysis and alerts via a monthly subscription. Company H signed a contract with Company I for a collection of specific reports and alerts for Company H's sales at RBR stores.

Initial results were promising. In one of the first "use cases" by Company H, Company I's alerts identified out-of-stock (OOS) conditions in stores that, when corrected, recovered an estimated $1.2 million in lost retail sales.

Company H has expanded the use of the application to some of its other retail customers. Many more "wins" have been scored by Company H, such as identifying and correcting distribution voids (missing products on shelves). While results have been good, Company H now wants to realize bigger benefits, create a self-sustaining process that delivers provable sales increases, and make a bigger impact on in-store replenishment practices.

**Key Points**

**Again, knowing about emerging software options yielded an attractive alternative.** Just a few years ago, companies that wanted to process and analyze retail sales data had to find a place to store the data, figure out how to "normalize" it to match the company's different definitions of sales, products and customers, and then determine what type of analytics software to use to make use of the data. This is commonly known as creating a data warehouse or a data mart, and a "BI stack," all of which are tremendously expensive and can take many months or even years to construct. Another option would have been massive Excel spreadsheets, which in fact Company H did use, but with limited results prior to acquiring the analytics from Company I.

**Company H had nailed a clear ROI.** It's hard not to overestimate the resources that would have been required to duplicate the analytics that were prepackaged in Company I's application. Such an effort would have taken a year or more just to put the data into a usable format, in a standard relational database, and to build or set up the analytics application(s). That's not including costs for data architects, programmers, database administrators and expert users who would know how to structure the data into usable reports and alerts. The cost of one good analyst alone at the time was more than the monthly cost of using Company I's application.

**Limiting the scope of the project to a narrow set of requirements made for a quick implementation.** The company decided that only three to four key indicators of retail sales conditions were essential for it to positively impact OOS through corrective measures. Yet there was much more data available – inventories, pending deliveries, pricing information – that the company chose to leave out of scope in order to get a quick win on its main objective.

# 4
# CASE STUDY SUMMARY

---

A quick review of what we've learned from the preceding case studies:

### Losers

As you saw in the preceding case studies, the money pit is easily found. Often you don't realize you're in it until it's too late. Most likely the people in the money pit case studies never thought – or at least never planned on – their project ending up in the dark hole. Had they seen how other, perhaps similar projects ended up failing, they might have avoided that fate. Here is what we learned about projects ending up in the money pit:

**Unclear ROI or payback based on unrealistic or false assumptions.** An unduly optimistic view of projected benefits. Vague promises of thousands of hours saved.

**A complex, unproven software solution.** A kind of hubris about the sophisticated design the team has developed. A belief that no matter the complexity, the software can handle it.

**Technology not easily assimilated by its intended users.** Often really smart people who understand technology don't realize that actual users won't have that same ability. User involvement from the start can prevent this, but among users there will always be varying skill levels. Keep it very simple.

**Unclear requirements or continually changing requirements.** Not getting into enough detail about exactly what you want the software to do in each business scenario. Allowing mid-course or late-course changes to requirements without evaluating their impact on the project.

**Inadequate vetting of the software leading to delays and extra costs.** Buying software based on demos and general high-level comparisons of what you need and what the application can provide. Remember, an application can do anything you want it to, as long as you have enough money.

**Big bang cutover without adequate risk management.** Not thinking through contingency plans if failures occur on or after launch date. Not willing to consider a phased approach because of artificially imposed deadlines.

## Winners

**Projects with an unambiguous and demonstrably positive** ROI based on realistic assumptions. Project ROI that is vetted and challenged, under different scenarios in case the application doesn't deliver all benefits as expected.

**Technology readily available in the market** and proven to work in other enterprises, with limited customization. Ideally a long list of satisfied customers, whose business models and needs are similar to yours. Software that attracts user groups and forums, and for which programming and other experts are available in the market to hire as support for your project.

**Full under-the-hood evaluation** of the fit of the system with the business. A set of pilot work sessions, with all participants present, where users step through each business scenario using the application.

**Narrow and specifically defined scope of implementation**; clear and unchanging business requirements. A strong defense against mission creep. Requirements that are documented in detail and that include all possible business situations and exceptions.

**A strong, experienced, focused team** with a mandate from top management and a simple, clear objective. A team made up of people who have been temporarily freed from day to day business to focus on the project. Ideally several team members who have worked on system projects before. No ambiguity about any part of their mission.

# 5

# THE HUMAN FACTOR

"And all the science I don't understand. It's just my job five days a week."

- 'Rocket Man,' by Elton John and Bernie Taupin, Uni and DJM Records, 1972

Betty Anne Graham stared at the touch screen mounted on the electrical cabinet. With a gloved finger she tried stabbing different diagrammatic views of the automated pallet storage and retrieval system (ASRS) to figure out what was wrong. Blowers mounted on the ceiling of the refrigerated warehouse circulated 37 degree air at 60,000 cubic feet per minute, emitting a constant roar that made it hard to hear anything except the ear-splitting klaxon alarm now going off. Betty shook her head and reached inside her stained and frayed thermal coveralls for the two-way radio. Her watch read 4:30 pm.

Until the fault was corrected, the entire assembly of conveyors, RFID readers, transfer stations, wrappers and labelers would be stopped, eventually causing a backup like a traffic jam all the way back through the plant to the production lines. The three phone-booth-sized automated cranes – which traveled smoothly along fixed rails like elevators moving horizontally and vertically inside the nine-story honeycomb of steel storage racks – had also stopped, meaning no product could be put away or shipped.

It hadn't always been this complicated. Betty made $18.50 an hour, plus overtime, had worked at the Really Swell Nutrition Co. (RSN) plant in Butte, Montana for 16 years, and was beginning to think about retirement; she'd be 61 in a few months. She and her husband, Don, had raised five daughters on a soybean and dairy cattle farm in the rural area south and west of Butte, in the shadow of the copper mines in the western hills where Don had worked for almost 30 years. She would rather be spending more time shopping with her girls and getting ready for her first grand-child, due in five months.

She was "forklift Betty" to her coworkers, an expert operator of the 7,000-pound forklift trucks used to shuttle the one-ton pallets of the company's product around the warehouse, lifting them 24 feet in the air and placing them in a rack location that measured four feet by four feet.

Betty's experience driving farm equipment gave her a leg up on the skills needed to drive a forklift truck. She had the finesse and hand-eye coordination necessary to gently tap or nudge the truck's hydraulics one way or another to move the load just the right number of inches up, down or sideways. She also followed to the letter – unlike some of her co-workers – the inventory control procedures critical to keeping track of the warehouse's receiving and shipping of 500 or so pallets of finished product.

On this particular day, Betty was tired of it all. The fault in the system she was staring at was all too familiar – it or a fault similar to it happened several times a day, as it had during seven weeks of testing before RSN switched over to the new system. From what she had heard, the company had spent $14 million building the new refrigerated warehouse and installing software systems that automated the movement of product from the production line to storage and then determined which pallets of product to retrieve from storage and deliver them to staging lanes for shipping.

There were three new software systems, all coming together to operate the automation in the warehouse, and no shortage of problems with each one. Everyone who worked in the warehouse was expected to be able to troubleshoot the systems, but training had been sparse. They were expected mainly to learn on the job.

Eddie Murkowski, a transplant from the New Jersey suburbs of Philadelphia, ambled up to Betty, hands in his pockets. He too looked at the touch-screen, tried tapping a few icons and shrugged. He was burned out from the long hours of testing and troubleshooting the new system. Just recently promoted to warehouse supervisor, Eddie hadn't anticipated working 12 to 14 hours a day just to keep the operation afloat. Eddie's boss, the young and energetic project lead, was the expert on all this stuff, but he tended to fix things just to get them done and consequently few people on his team learned the essentials they needed to know to operate the system.

The fault the touch-screen displayed said "Contour fault at get location 05355. Restart delay failed." Neither Betty nor Eddie had seen this fault before. Betty had tried to radio a maintenance technician, but her unit was malfunctioning and full of static, so she started for the shipping office. On the way she noticed that one of the Automated Guided Vehicles (AGVs) had stopped and was emitting an alarm and a flashing yellow light. She might know how to fix the AGV, but didn't have time at the moment. She also noticed three pallets had just been rejected by the put-away system and were waiting in the rejection lane for someone to diagnose the problem and try re-inducting them into the ASRS.

Thirty-one thousand feet over eastern Tennessee, 29-year-old Really Complicated Technologies (RCT) software engineer Mike Lipinski opened his laptop and connected to the in-flight Wi-Fi.

RCT had installed the control software for RSN's ASRS. A text message from his boss pinged on his screen: "RSN is down. Can you diagnose? Contour fault." Mike had told them a million times how to fix a contour fault, which resulted when part of the product or pallet interfered with the photo eyes that tracked pallet movement and signaled conveyors and lift platforms to move or stop.

Logging into the same screen Betty was staring at, he saw that it wasn't just the contour fault; someone had fixed that fault and tried the restart procedure, but that had failed. Hmm.

Back at the plant in Butte, now 45 minutes into the stoppage, the maintenance technician had arrived, and watched the screen as Mike moved the cursor around from his airplane seat. At about the same time, Betty was summoned to production line three, which would have to shut down if someone didn't remove product from the end of the production line.

They had prepared for this scenario – ASRS down for more than half an hour – called a "business continuity plan" (BCP). The BCP involved a forklift operator attaching a bar code label to the pallet at the end of the production line and scanning it into the system. This would remove pallets from the production line and allow the line to continue running, because the AGVs that normally did that were backed up like cars on a freeway, waiting for the ASRS to start up again. But they could only do this for so long; with only two forklift operators working the shift they couldn't possibly keep up with the plant's six production lines all running at the same time.

Really Swell Nutrition Co. had started out as a milk processing plant, and then later began manufacturing milk powders and supplements. Its newer products were higher-margin condensed and fortified protein mixtures in pill, powder and liquid form. It had a solid market share, and its newer products were in high demand; sometimes the plant could not produce all of what customers were ordering. Which meant it was a very bad time to interrupt production.

By the time Mike Lipinski's flight was descending into the Northwest Arkansas Regional Airport, he still had not cracked the root cause of RSN's ASRS restart fault, but he had a theory: somehow the power to the on-site server that ran the ASRS at RSN had been interrupted, and the server didn't restart correctly. He was in the process of emailing the plant's IT manager when the plane dropped below 10,000 feet and his Wi-Fi connection was lost. At about that time – 6:30 pm Mountain Standard Time – production lines 2, 3, and 5 at the plant shut down. There was no more floor space for any pallets, and one of the forklift operators had left early.

RSN's IT manager, 44-year-old Chris Lonegan, slid off the Six Shooter ski lift at the Moonlight Basin Ski Resort, ready for another run, but first he took in the incredible view of the surroundings, made possible by the invention of night skiing. Living and working in Butte had made Chris a pretty good skier, and he took advantage of his proximity to the slopes whenever he could. He and his wife didn't have kids and were free spirits enjoying the great outdoors and building a rental property business on the side. He had just pushed off for another run down Highline Trail when his cell phone buzzed in his back pocket. Whoever it was would have to wait.

Scott Segretti, RSN's Butte plant manager, wanted answers. He was responsible for the building that included the warehouse, but not for how it ran – that belonged to Supply Chain, a different functional silo in the company. He noticed that the project lead, who was also the warehouse manager, was nowhere to be found at the moment. The ASRS stoppage had eventually stopped all six production lines; they had been idle for two hours now.

Pretty soon Segretti would have to order raw materials that had been prepped and batched to be thrown away – they could only be held in their raw state for so long before beginning to spoil. Segretti's boss, the VP of manufacturing, expected operational efficiency (OE) above all else – and what Scott saw around him was a big hit to his OE caused by someone else and with no clue when it would be resolved. He pulled out his cell phone and dialed.

Alex Abrams had come to Butte two years earlier from the company's Houston plant, and had performed well. He had just turned 30, was single, and like the plant's IT manager, loved the Big Sky country in Montana. He was very quick with new systems, and was a logical choice to lead the implementation of the ASRS project. In the middle of the project, his boss left for a new job in California, and management promoted him into his previous boss' position as warehouse manager.

But the project had taken its toll: Alex had few people on his team with strong system skills, the ASRS control software proved very unpredictable, the launch was delayed three weeks to perform more testing, and now three weeks after launch it all seemed to be falling apart, with problem number one being the quixotic ASRS control software installed and managed by Really Complicated Technologies. To Alex, it also seemed that as time went on, his people got weaker; inventory was a mess, people were making simple mistakes, and his recently promoted supervisor seemed overwhelmed.

Alex saw the name on the caller ID and rolled his eyes. He had spent 30 of the last 48 hours in the warehouse, fighting problems as they arose, going from one to the next. It would be another long 48 hours. Scott was his usual demanding self, asking first why Alex wasn't at the plant, then taking him to task for the sudden shutdown. Segretti wanted a fix now, and hung up with Alex so he could phone Alex's boss.

Chris Lonegan had gotten off the slopes and went out with friends for a few beers. It was 11:00 pm before he checked his messages. He tried to call Mike Lipinski back, but got voice mail. He went into the plant and diagnosed the server, seeing no evidence of a previous power interruption. He and Abrams, Murkowski and Segretti huddled and tried calling Mike Lipinski or anyone at RCT. The warehouse team had gone into complete manual mode, trying to store inventory and pick customer orders without the ASRS control software functioning.

By 7:30 the next morning – Thursday, one of the plant's busiest shipping days – tension and frustration filled the air. RSN had already missed four important customer shipments – that business was gone because customers would fill their orders from competitors. A truckload order to one of PNC's big customers was worth about $105,000 in sales. So far the software problem had cost $450,000 in lost sales.

––––––––––––––––

The preceding story is true, but names, locations and some figures have been changed to respect privacy. I spent four months on this project during its latter half, and had plenty of time to catalog what went wrong, and how we as a team could have prevented some of the failures. They are presented here because I think they are also broadly applicable to any significant investment in technology. Lessons learned = smarter decisions the next time.

Epilogue: The root cause of the failure was a change in the code of the ASRS control software that had been done earlier in the day by another member of the RCT software team, David Waring. David had been fixing one of the outstanding issues on our punch list, and had inadvertently reconfigured a subroutine that functioned to automatically restart the cranes after a contour fault had been cleared. This was discovered around noon Wednesday and by 1:00 pm the ASRS was back up and running, but not before RSNC had missed another three shipments.

### Key Points

**Not everyone has the system skills you expect**. RSN was introducing more advanced technology than the warehouse team was used to. Nearly everyone except Abrams, the project lead, struggled with aspects of the new systems. Inventory became scrambled by people confusing an RFID number with an SSCC (or pallet ID) number, by people not linking the RFID number to the proper SSCC number, and by people switching the RFID-embedded pallet without scanning in the new RFID number. Basic understanding of how the system worked end to end was missing.

**People have their own lives and aren't necessarily going to sacrifice them for the success of your project**. By the end of RSN's ASRS project, a lot of people were tired, fed up, burned out. One of the team members said to me five weeks after the launch: "I am now going to take my life back." He didn't care anymore if his boss criticized his performance. He didn't care if the company fired him.

I think a lot of the stress on people was self-induced by our unwillingness to staff a strong implementation team, but there is a broader message: not everyone gets their jollies from a high-intensity technology project: they just aren't as interested as you and me, and like Betty, Alex, and Chris, they want to have a job that doesn't intrude on their time with family or with their interests outside of work.

**Classroom training is insufficient**. We had plenty of hands-on sessions with operators, shift leads, and supervisors, but not everyone attended these, and the training itself was mostly step-wise procedures using specific screens, not a deep overview of how things worked end to end. The ASRS faults came at the warehouse team rapidly and frequently. If the right person was standing near the equipment when the fault occurred, it could be quickly corrected. But often there was no one nearby because everyone was busy doing something else.

**The business has to staff up**. We had to carry out the entire project with the same people who were responsible for day-to-day operations in the warehouse. Guess which priority came first? The day-to-day operations always had priority, of course.

Consequently, a lot of people missed really learning the system because they were too busy doing their regular jobs. We needed extra people for maybe 12 to 16 weeks, but there was no budget for that, which brings me to one of my pet peeves: skimping on project resources only to spend way more than what you saved later on through business disruption, overtime, product losses, expedited transportation, or other expensive corrective measures. We kept the project team lean, but I think eventually paid the price for doing that.

**Relying on external support for your critical systems has its risks**. Not every company has the resources to develop and maintain its own ASRS control software; RSN certainly did not. This software sends instructions to programmable logic controllers (PLCs), which then start and stop machinery, open valves, run conveyors, etc. Many enterprises rely on firms like RCT to install and service these systems. RCT had spent months on site with us, and Mike Lipinski was a particularly dedicated soul, even in the darkest moments of our testing when it seemed like nothing would work.

What would I have done differently? 1) Hired temporary warehouse operators to run the day-to-day business while dedicating RSN employees full time to the project; 2) Negotiated on-site RCT support 24x7 for six months post go live; 3) Certified through training two RSN employees (IT technical people) to manage outage incidents with the ASRS control software; and 3) Secured a service-level agreement with RCT ensuring one- to two-hour maximum response time and possible penalties for excessive downtime.

**Be realistic about your backup plans**. The "highway" that transported RSN's finished product from production line to storage in the Big Box (what we called the nine-story set of storage racks) by the ASRS was only long enough to hold so many pallets before traffic would back up – about 40 minutes until the path would be jammed. The AGVs were designed to sense when another AGV was in its way, and would automatically stop if its path was blocked. But 40 minutes is a very short window within which to diagnose and fix a problem. Unrealistically, we had thought that any issue beyond 40 minutes could be managed by manually picking up pallets from the production line with forklifts and entering the proper data in the system using handheld RF scanners. With no extra forklift operators, however, this was an infeasible scenario when more than three production lines were running. Because of our human tendency to believe that things will work out somehow, we put more faith than was warranted in our risk mitigation plan.

**Spend a lot of time in your future state before committing to your investment.** It's important to step into the future to imagine how your technology solution will function, interactively with your people, to produce the benefits you want. Details are important in this exercise. I don't think our RSN team spent enough time imagining the future. If we had, we might have seen that 1) ASRS faults are going to be frequent and we need a fast response in all cases, around the clock; 2) there will be new work created that someone will have to do, such as re-inducting rejected pallets and daily inventory reconciliation between the three systems, and we hadn't planned anyone other than the existing team to do them; and 3) the ASRS control software isn't easy to learn and it will take a long time – much longer than we think – for most of the team to be self-sufficient with it. Maybe because of the pressure to hit the launch date we didn't have sufficient time to dwell on the future, but that situation did have its costs.

# 6
# NIP THE LOSERS IN THE BUD

"Sometimes your best investments are the ones you don't make."

- Donald Trump

This chapter focuses on that point in time when your enterprise is considering one or more investments in new software. It is in these gestational moments that you have an opportunity to shape the smart ideas and question the dubious ones.

How do software projects get started anyway? Someone went to a conference. A department manager wants to "streamline workflow." A sales VP says his team is wasting time with old and slow systems. A "transformational" project is launched. A new plant or warehouse is being planned. Customers start asking for things your company cannot do with existing systems.

In each case a person or group of people claims that, with a new system, all kinds of benefits are possible. But anyone can create a list of benefits; with some creativity you could gin up an attractive ROI for a ham sandwich.

Some project failures can be tied back directly to the original investment decision. Consider Denver Airport's initial decision: a state-of-the-art, complex solution, to be implemented in a shorter time frame than most run-of-the-mill baggage handling solutions. Author David Yardley, in his excellent 2002 book Successful IT Project Delivery: Learning the Lessons of Project Failure, describes the quite complete failure of the system upgrade attempted in 1992 by the London Ambulance Service (LAS). The disaster can be traced in part back to one of the original decisions, to select a small, little-known software firm because its bid was the lowest. Similarly, there was Nike's decision to include all of its products and suppliers in the initial launch of its new supply chain software.

In each of these cases, the main reasons for failure could have been nipped in the bud at the very beginning.

Notice also that these original decisions aren't just about the software selected; they are about many aspects of the project – the team chosen, the timeline, the way in which the software will be used to produce benefits.

First, some straightforward advice on when new systems make sense and when they don't:

## Good Reasons To Invest

- You can identify hard savings as a result, in labor cost, time to market, inventory reduction, or some other benefit, and the ROI is clear, substantial, provable and realistic.

- Your business has grown and you can no longer manage the increasing number of transactions without significant manual effort.

- You are having a hard time with internal financial controls, and/or with meeting regulatory or taxing authority requirements for record keeping and internal checks and balances.

- Through acquisitions of other companies you are now faced with managing the enterprise via different software platforms; integration is too complex and expensive, and standardizing on one platform allows better visibility and control of the business.

- Your customers and/or suppliers want closer collaboration with your company on business planning and execution, and to do so will require some customization of software, new or enhanced e-commerce connections, master data linkages, and cross-company sharing of demand forecasting that cannot be accomplished using existing systems.

- Your main customers are now requiring data or services your systems can't provide, and this is putting your company at a competitive disadvantage.

- You are now a $1 billion company and realize you are managing most of the enterprise with Excel spreadsheets and home-grown applications, with code and database schema that is not documented, and whose insides are well-known by just two people in the company, both of whom could retire, die or quit at any time.

- The version of your packaged application is due to "sunset" and your vendor's upgrade is expensive, or you are questioning the continued value of the application, justifying a look at other alternatives.

## Bad Reasons To Invest

- The technology demos well and looks cool, but the ROI is unclear and seems contrived (the shiny object). The new system would make life easier for many, and reduce non-value-added work, but not necessarily result in lower labor costs. People complain about the extra steps and wasted time with the current system. In this case, more homework is needed to see what additional leverage the new system would bring. Is the company growing fast? Could it reduce future labor costs? Are there customer demands not currently being met that could be satisfied with the new solution?

- A new executive is from a company that runs SAP and believes this platform is the best solution for your company. Nothing wrong with having an opinion like that, but is your company ready for the precision and business process discipline required by SAP or any other ERP system? Moreover, is there a business case for replacing what you have? SAP is not the only enterprise software available, and in many functional areas outside SAP's core capability there are more robust and less expensive solutions available.

- One of the departments in your company has gone off on its own in pursuit of a new system. Having spent weeks evaluating software demos by different providers, your VP of Sales and her team is convinced the company should purchase and implement vendor B's solution. She has also gotten the support of the VP of marketing. These actions alone are not sufficient to justify the investment. Validate the ROI if there is any, and confirm that the company has the time and people resources needed to complete the project.

- The 700 employees in your organization are becoming increasingly frustrated with their buggy, quixotic and slow-to-boot Windows-based PCs, and want Macs or iPads instead. Some senior managers have purchased their own Macs and had them configured to work in your department's network. A complete desktop switch would mean having to also change or reconfigure some of your enterprise applications. The answer here is not so easy – despite the overwhelming popularity of MacBooks and iPads. Many enterprises have invested in applications and help desk support systems and people that are geared toward Windows-based PCs, so it's not just a matter of buying everyone an Apple machine. The decision should be based on an analysis that takes into account total cost of ownership between the two types of PCs, including the costs of maintenance, troubleshooting, and personal productivity.

As Donald Trump said, some of his best investments were those he didn't make. Investing is emotional. As a real estate investor I have looked at many properties, and with each one, despite the hardened skepticism I naturally apply to all of them, I also apply a similar but opposing thought: how could this property be changed, developed or used differently to produce value? The same is true with IT investing: there is always a possibility out there, imagined by someone, that a particular technology investment could produce different processes, eliminate waste, inspire minds, please customers.

## The Shiny Object

You've seen it before – how projects gain institutional momentum even though the hard reasons for it aren't clear.

An expensive multi-year project is hatched by an executive's desire to enhance his or her career, or from the need for some level of management somewhere, to convince higher executives or shareholders that *action is being taken to move the enterprise forward* by technology-enabling new and improved processes.

Ideas – the seeds of projects – have a way of gaining momentum without the right level of rigorous questioning and unbiased analysis. Notice how a project, slim as the specifics might be, gets an impressive name, such as Phoenix, Gemini or Athena, and then almost automatically has credence as a legitimate, important initiative that must be done. Once the project gains notoriety in this way, there is a bias to produce something from it, and often that bias to do something ends up being the purchase of a new system or systems.

Years ago I worked for a company that announced a project to "transform" the company chiefly by "integrating" different departments; we'll call this project Excalibur. A team was formed, made up of the best and brightest, from a cross-section of functional departments. A lot of full-day meetings were held. I wasn't involved in Excalibur, yet, so I didn't know what the meetings were about. But I do know that the first thing – no, make that the only thing – the Excalibur team set out to do was to evaluate packaged software solutions...for what, it wasn't immediately clear. But it occurred to me that the team was completely bypassing important steps, like stating a problem and business goals, identifying broken processes or dysfunctional parts of the organization, and defining a desired future state.

Epilogue: the project took nearly five years to complete, and resulted in replacement of four legacy applications with a new, interconnected ERP system. Two years after the completion of Excalibur, a new project was launched to replace everything with SAP.

You have to wonder how enterprises whose software projects failed would have fared had they done nothing. In many cases the answer would be obvious: they would have been much better off doing nothing. Refusing to participate in a high-risk endeavor is a smart move, but in the 21st century that path is not always realistic. Growing companies cannot use calculators and spreadsheets forever. Regulatory requirements, billing, payroll, invoicing, accounting, tax filing, and many other activities just can't be done anymore in a manual fashion; the administrative cost alone would sink almost any business.

Anyway, let's say the bandwagon for a software overhaul is gaining momentum, and it appears that key people in your enterprise are locked on to their arguments to do something to improve the software tools they are currently using. You, wanting to make the smartest possible decision and dodge the abyss of cost overruns and business disruption, take notice, and begin to cautiously participate in the decision process.

This stage – the birth of a project – is hugely important. From here can spring the beginnings of a slam dunk investment or a money-losing disaster.

How do you make the best use of this moment in time?

1.  Position yourself as a gatekeeper of information, declaring what information is known and what additional information is needed before decisions are possible.

2.  Evaluate the assumptions behind the interest in a new system: are they aligned with your company's objectives, and are they based on measurable business impact? You will find that everyone has different assumptions: some think the problem is X, some think it is Y; and many assume that the future application will be much better than what they currently have.

3.  Show people how the dots have to be connected for the whole thing to make sense.

## You As Gatekeeper

As a gatekeeper of information, one of your main roles is to *keep asking questions*. This has the effect of exposing and testing the initial rationale for a project. For example, in the scenario below let's say you are the CIO or CEO.

CIO/CEO: What exactly are the benefits of this investment?

Manager: We'll be able to cut customer order lead time and reduce our on-hand inventory.

CIO/CEO: Great. How?

Manager: The new system will give us real-time visibility of our vendor inventories and plant inventories, and instead of waiting for reports we'll see our inventory positions and planned production and receipts real-time.

CIO/CEO: So people will be monitoring inventories, planned and actual production 24x7?

Manager: Well, that's possible, but might not be necessary...

CIO/CEO: How exactly will the order fulfillment and material buying change after the new system is put in?

Manager: As I said, we'll be able to see the real-time situation, and be able to make better decisions...

CIO/CEO: Yes, but exactly what will change, in terms of process, compared to today, to give us these benefits?

Manager: We'll be making smarter decisions because of the real-time information and ....

Does this dialogue ring true in terms of how projects are justified? The problem is that the manager hasn't thought beyond the basic headline argument that *real- time views will make everything better.* This should be a danger signal – people have bought into an imagined benefit without proving out to themselves exactly how this benefit will be achieved.

Some other relevant questions in this scenario would be: "Is lack of real-time visibility the only constraint to lower inventories... how will lower inventories translate into real savings besides reducing cash flow...how will you change your decisions about production and will the company be able to execute these changes in production scheduling...when customers miss the order cutoff do they order anyway and take delivery later...?

Asking questions, especially specific ones, quickly changes the conversation from one of vague potential to real-world feasibility.

## Evaluate The Assumptions

Behind every idea is a set of assumptions that are usually exposed simply by asking "why"? It's your job to test these assumptions, as part of winnowing out the losing software propositions or by making them more viable as sound investments. Sometimes these assumptions are wrong – and a lot of them need to be right in order for a project to succeed.

Many people don't realize the number of assumptions they make when a technology project is launched. Among them: what they saw in the demo or pilot will work in the real world; the software will meet all the business requirements that were specified before the project started; the team won't have to make any customizations other than what was already identified; users will quickly learn and accept the new system; the project will be completed on the promised date.

In the dialogue example between the manager and the CIO/CEO, recall the manager's statement: "The new system will give us real-time visibility of our vendor inventories and plant inventories, and instead of waiting for reports we'll see our inventory positions and planned production and receipts real-time." Taken at face value, this statement implies acceptance of the following assumptions:

1. The way we think of "real-time visibility" of inventories, production and receipts is the same as what the system can provide.

2. The view of said data will be in a useful format and will provide all the data we need to make better/faster decisions.

3. These better/faster decisions will enable us to let our customers order within a shorter lead- time window and will reduce our on-hand inventories.

4. The savings from lower inventories and the additional sales from our late-order customers will more than pay for the cost of this new system.

5. A change in business process (i.e., how we manage inventories and production) would not produce these same benefits.

6. Out of all of the possible system solutions this one is the best choice from an IT strategy, cost and ongoing support standpoint.

## Connect The Dots

Too many people rely on software as the answer. "If we only had a flexible system we could …" But software alone is never the answer – software is just a set of tools. Tools have to be used to produce anything useful. And they have to be used in the right way, by those who know what to do with them.

The same is true for dodging the money pit. If the dots are connected right, you have a win: fixing the right problems with the right tools used in the right combination and manner by the right people, yields something valuable.

So the logical question is: what are you going to fix, build and maintain with the tools in the toolbox? How long should it take? What materials will be used, who will use the tools, and how will you know when you're successful? You, as gatekeeper, pose these valid questions in order to help you and your enterprise connect the dots and illustrate the right solution.

In Case Study #4, Company X bought the tools but didn't really change the (customer ordering) process. The dots weren't connected and the benefits of the project were never fully realized. In Case Study #5, the Census Bureau didn't connect the right people skills, the right process, or the right training with the new hand-held devices.

Make people understand that it's a combination of process, tools and people that will produce the win. *If all the dots aren't connected, it's not a win.*

# 7

# STEP FIRMLY INTO THE FUTURE

"The best way to predict the future is to create it"

- Abraham Lincoln

A true win in the world of software investing is planned, known and secured from the beginning because all the right pieces are put into place. Software investments shouldn't be like buying stocks. You aren't speculating here; the ROI should not be a guess or an estimate or a hope, or a fantasy everyone has bought into.

In a race, you don't jump off the starting line, then figure out how you're going to get around the track 10 times, manage the hurdles, and come in under your time objective – you see this beforehand, in your head and in your hours of training for the race. The same is true here: you must step into the future and define in detail the system-process-people combination that will bring the financial returns you expect.

Not spending sufficient time defining the future has torpedoed many projects.

The Denver Airport team designed tracks for baggage carts to travel on with angles too sharp for the speed the carts would be going; carts fell over during testing. Some basic physics calculations would have revealed that their plans for speed, baggage weight and turn angle were incompatible with one another. They had defined the future, but it was unrealistic.

## Keep The Scope Manageable

"Scope" is simply the number of business processes that the systems project encompasses. An over-ambitious scope is one way to increase the odds of failure. Nike launched its new system with thousands of suppliers and distributors all at the same time (scope too big).

How do you find the right scope? Determine *which areas of the business would benefit the most* from a new or better application. Can you define the specific problems that are leading your enterprise to consider new software? Where are those problems located – in what functional areas and related to which current (legacy) system? Is the problem that a) a particular application is too limiting; b) a group of applications are islands and that integration of them would yield benefits; c) none of your applications are integrated; or d) something else?

**Consider a range of scope options to find the optimal one**. In some cases, expanding the scope of a new application beyond "problem areas" can be the optimal choice. The process is iterative, and you should consider several alternatives. For example, implementing a new accounting system may satisfy most of a company's needs and produce a good ROI on its own. But expanding the application footprint to, say, payroll and purchasing, may result in an even better return because it simplifies integration costs, eliminates more manual work, and may strategically be a better decision.

**Set up a framework to evaluate each scope alternative**. In a framework you can evaluate each scope option according to such factors as cost, complexity, length of time to implement, risk to the business, ROI, required internal resources and strategic value. Then you have a logical basis for your decision.

**The scope of an ERP project does not have to be huge and overwhelming**. You can be selective in what processes to migrate to an ERP system, and you don't have to convert everything at once – both of these steps will reduce the overall risk of the project. As mentioned earlier in one of the case studies, companies can implement demand planning systems first to shake out the bugs in what is traditionally a complex and parameter-sensitive application. The core financial systems of an ERP can also be phased in first before everything else.

## Draw The Future In Detail

Company P is a firm I have worked with that helps companies to achieve ambitious goals. One of the firm's methods is to engage in a "Merlin" exercise – so named for the legendary magician who claimed to come from the future. In this exercise, groups describe in words or illustrations the future of their company in three, five or more years, pretending that ambitious goals have already been achieved. How did the company get there? What do processes look like? The more detail, the better. How did we get rid of all that non-value-added work? How did we double business with the same number of people? This is the approach you need to define how your software investment is going to change your business results.

**Ignore the systems for now**. For now, don't try to imagine your current system with great new features. That is too limiting. For this part of the project, keep "the system" generic. It can and should appear in your future plans, with specific capabilities, but not as a specific system. When you have selected specific software and become aware of its capabilities and limitations, come back and re-draw the future processes *as they will be executed with that particular application.*

**Define what problems are fixed in the future state**. The present must have some characteristics that people want to change, or else there would be no interest in a new system investment. So start with a few *problem statements* to articulate why you want the future to be different from the present.

A useful problem statement sounds like this: "While visiting customers, the sales team cannot access order history nor place additional sales orders, resulting in lost revenue to the company."

Why is it useful? Because it is specific in stating what cannot be done today and what the business impact of that is.

A deficient problem statement sounds like this: "The company lacks a robust set of applications for the sales team."

Why is it deficient? Because it doesn't say what specifically is missing, doesn't define "robust," and doesn't identify why this situation is important to business results.

**Example: Johnny's Pizza.** Johnny is computerizing his business, after years of running on pencil and paper and Excel. His problem statement is: "Johnny's customers can't order online and so the shop loses business to other shops that have online ordering. Customers have to order in person or by phone, transcribing mistakes are made, and during busy times customers have to wait on hold. Johnny's could sell more pizza more efficiently if customers could order and pay online."

Johnny's team starts to describe the future, in detail:

1. The customer will access www.johnnyspizza.com and will log in as a returning customer, register and log in as a new customer, or order without registering.

2. The registration process will create a username and password chosen by the customer and will request from the

customer the customer's name, delivery address, payment method, and phone number. This information will be displayed each time the customer returns to the site and logs in.

3. The customer will see the full menu in a way that lets them build their order by selecting size, type of dough, toppings and extras.

4. Prices are displayed for each choice. Subtotals are displayed as the customer chooses items from the menu. A final total is displayed for the order.

5. The customer is asked if their order is final Y/N. A "pay now" message is displayed along with a form to fill out if the customer is not registered. Returning users will view a confirmation page on which they will confirm all information including form of payment. A randomly generated order number is also displayed.

6. All orders are displayed to Johnny's team in the order they were received, with the time they were received. Specifically, the size, type of crust, toppings and extras are shown, the amount due if not paid online, the name and address of the customer, and so on.

Your enterprise probably has more complexity than a pizza shop, but the point is the same: *A good definition of the future goes into many details.*

## Key Steps In Defining The Future

### Preparation

Get your business and subject matter experts and key managers or other stakeholders in a room, focused for a day or more. Outline the agenda: "The purpose of our meeting is threefold: 1) to define as a group exactly those business problems we wish to address and how they impact our business; 2) Imagine what the future looks like in as much detail as we can, including how specific business processes would change; and 3) quantify the changes we seek in our business – costs, sales, headcount, customer service, and so on."

At this point you are not defining what the software or other technology must do. The job of technology is not to simply exist on its own; it's to enable your efficient and profitable business processes.

This is a highly interactive session best executed with visuals where possible – whiteboards with illustrations, lists, concepts, etc. I once participated in a similar visioning exercise, where groups were asked to draw the future without words. The artists won the day, but the idea was to paint a picture so vivid it instantly captured the future state you desired, and was clearly understood by everyone else in the room.

### State the Problem

1. What is wrong or suboptimal – what is broken, what capabilities are lacking, or what opportunities exist that you can't go after because something is missing? "Customers want our products re-packaged into special formats to sell in their stores as unique offerings…our systems aren't set up to manage orders and invoices for this re-packaged product."

2. What is the impact of the problem? Who does it affect – customer, supplier, employees; how does it affect performance – service, costs, efficiency; and what is the lost opportunity – sales, market share, business growth?

3. In each case be specific. "We need more flexible systems for managing customer orders" is not specific. "We cannot change anything on a customer order without deleting and re-entering it" is specific.

### Imagine

1. When the problems identified in the first step are resolved, what does that look like? What would it look like if, for

example, the problem of having to delete and re-enter a customer order just to change it were fixed? "Service reps can retrieve from the system a customer order, go into an edit process, make any change required, including products ordered, quantities, delivery dates, method of payment, and shipping options, then save the order."

2.  Don't be constrained by today's business situation, type of system architecture, or even the preferences of senior management. Don't limit your thinking to just small problems. Imagine a bigger picture, years into the future, when the company is transformed from what it is today...what would that look like?

3.  Start with the big picture – "Sales have doubled" – and work downward to specifics from there. "We started selling through a web portal." The big picture is the eventual outcome you want, the specifics – a web portal, for example – will be the things around which new business processes will be formed. The new processes are what you want the new technology to enable.

4.  Define the new processes. A 'process' is just the sequential steps that lead to a particular outcome. The easiest way to define a process is to ask questions. What is done first? Then what? Then what after that? You may end up with three new processes, or 20, depending on how big your project is.

**Quantify & Specify**

1.  Determine what the future state is worth to your enterprise. What costs do you save, what new customers do you gain, what new products or services can you now sell? Here you need real dollars, not just imagined benefits and rough guesses. Also key performance indicators such as cost per pound, customer order fill rate, and % obsolescence and write-offs, for example. Those numbers that you use to track the health of your business.

2. Zoom in on each part or step in your future process and list the requirements of the new software. Example: "enable lookup of a customer's last order," or "display the location of the customer on a map using street address, city and zip code."

3. Log requirements into a spreadsheet file and track the progression of each throughout the project (open, in progress, enabled, and "gap" – the software does not meet that specific requirement).

4. Use the requirements file to evaluate alternative system solutions. It's easy to calculate a "% fit" using your requirements list – if you have 200 requirements (not unusual) and the software appears to meet (really meet) 165 of those, you have an 83% fit.

# 8
# HAVE A
# STRATEGY!

---

If you don't know where you are going, you will wind up somewhere else.

- Yogi Berra

## What's A Strategy?

Before leaping into a software project, determine *what your company's general approach is* to making these kinds of decisions. What does a strategy software or IT strategy look like? Here are some examples:

**The complete outsourcing strategy**: The company outsources all IT and software needs. It does not own any applications; it just uses them as a service. It pays these service providers to upgrade or further customize the applications it uses on an as-needed basis. Users have full access to service providers for day-to-day support, but the company also has both business and IT professionals who are experts on these systems who can provide internal support.

**The pure ROI-based strategy**: The company bases its decisions on software investments according to strict financial hurdles. Every investment has to pay back within, say, 36 months, and it includes all external and internal costs in that ROI model. All projects are reviewed post-implementation to confirm the expected ROI was actually achieved.

**The anticipated growth strategy**: The company uses ROI as one measure of whether the investment should be made, but it also heavily weighs anticipated future needs. It looks ahead a few years to see what customers, suppliers and competitors might be doing. It might end up making a new application investment for purely strategic reasons.

**The complete in-house strategy**: The company's business model is so unique, and its processes and competitive advantage so specialized, that it never finds what it needs in commercially available software packages, other than software for basic support functions like HR, payroll and accounting. It uses software programming languages that are well-known and well-established in the industry so that it has a large pool of talent available for its growing needs. For this company, software is a strategic necessity and an area in which it invests heavily.

## Make The Strategy Fit Your Business

I could go on, but you get the basic idea: every enterprise has some kind of software investment strategy. There is no right strategy, but here are a few things I think are important:

**Align your software selections with who your company is and where it's going**.

- A stable, 30-employee company with no significant growth plans can do just fine with many off-the-shelf applications or one of the many SaaS providers now coming onto the market.

- A vertically integrated auto parts maker with 3,500 SKUs and contracts with large auto companies probably needs top-of-the-line supply chain and distribution applications capable of strong integration with supplier and customer systems.

- A credit card company with 4 million customers needs very secure applications with redundant backup capabilities and rich data mining software to analyze customer transactions.

If your strategy is to outsource some functions in two to three years, you might not be interested in spending money on new applications for those functions. Many companies outsource warehousing and distribution, so they never have a need to invest in those types of applications. Or, your company may be growing and large enough to consider combining operations from separate divisions under one application umbrella, in which case you might establish a data center and a common enterprise core application, such as SAP or Oracle.

**Invest in applications that help you do excellently those things that are critical to your business**. Go light on or buy generic applications for functions that you consider support, and not central to your business strengths, unless you see a clear, hard ROI for one of those areas. But get or develop the best possible applications for those functions that make or break the success of your business. A manufacturing company will want applications that support "lean" (highly efficient) manufacturing processes, because the efficiency of the firm's manufacturing processes will be one of the biggest contributors to profitability.

**The packaged application options available to you depend on what part of your enterprise will use the application, and what you want the software to do**. If you are a large health insurance company with millions of customers, you will likely find the most suitable products to be those tailored specifically for the insurance industry. Your system may have to process millions of transactions a month, and keep detailed records on each one of your customers. This type of application will be completely different from software used by a manufacturer of packaging materials that are sold wholesale.

**Completely custom software can be part of your strategy if what you need is unique to the core success of your business.** Otherwise, look for options in commercially available software. Building an application from scratch is a very expensive endeavor, for two reasons: 1) there is a tendency to continually expand what the software can do, thereby making its cost open-ended (remember the adage: "programmers never finish a program, they just stop working on it"); 2) a good portion of what you want the software to do may already exist in an application readily available on the market. In writing the application from scratch you are simply reinventing what someone else has already done.

**There is no general rule of thumb for these decisions, just common sense.** A company with $10 million in revenue simply won't be able to afford an application that will cost $2 million to launch and another $500,000 a year to support. The logic of aligning your software tools with who you are as an enterprise is just one more ingredient in staying away from the money pit.

## Tread Carefully With Custom Software

The overhaul of North Carolina's Medicaid claims system, still under way after four years, is relying on custom-built software. All 50 states process Medicaid claims. Several software applications are available for medical claim processing, posing the question of why the state chose to build its own solution. The North Carolina official in charge of the project, when asked by a reporter why costs had so overwhelmingly exceeded the original projection, explained that changing requirements kept adding programming work. Certainly if it takes years to complete a project, new requests for functionality are bound to arise; if these requirements are accepted into the project the endeavor can go on forever – a cycle that feeds on itself.

For custom software projects, vendors generally bill on a time-and-materials basis, a fixed fee, or a combination of both. Custom software developers run the gamut from large companies with hundreds of staff to one-person programmers. Custom developers, whether large or small, are most useful for discrete projects where a tool must be built, a functionality extended, or integration between two systems developed. These skills can be useful whether you are writing your own code or modifying a packaged application.

**Understand what you are getting into with custom development.** In *The Mythical Man-Month*, computer programming legend Frederick P. Brooks likens software development to the struggles of prehistoric beasts in the tar pits: "Large system programming has over the past decade been such a tar pit, and many great and powerful beasts have thrashed violently in it. Most have emerged with running systems – few have met goals, schedules, and budgets. Large and small, massive or wiry, team after team has become entangled in the tar. No one thing seems to cause the difficulty – any particular paw can be pulled away. But the accumulation of simultaneous and interacting factors brings slower and slower motion. Everyone seems to have been surprised by the stickiness of the problem, and it is hard to discern the nature of it."

**Custom projects are especially hard to estimate and manage.** Brooks argues that software development projects often fail to meet goals because:

- Estimating techniques are poor, and reflect the pervasive optimism among programmers that everything will proceed according to plan.

- Using estimated man-months to determine cost and length of a project is a "dangerous and deceptive myth, for it presumes that men and months are interchangeable."

- Project status monitoring is not rigorous enough to prevent slippage, and when projects miss milestones, the natural tendency is to add more people, which, counter intuitively, will delay the project even more.

**Some small parts of a project are good candidates for custom development.** Integrating two or more applications, for example, via custom data transfer mechanisms or EDI or some other way, is an activity well-suited for custom development. So are data marts or customized reports that extract data from the main application and present it in a specialized way.

## Know What Your Choices Are

In the 1990s – boom years for enterprise software – men and women in suits came to your office and handed out color presentations, with your company's logo on the front, next to their company's logo. There were colorful pages of boxes and arrows, pyramids, and images of speed and agility. There were pictures of wheat fields, race cars, spacecraft, planets and astronauts; metaphors for your business challenges (and triumphs made possible by their software). There were slides that implied huge increases in productivity and profit, as well as your company's ability, with their product, to beat the competition. Perhaps a demo was put together.

Next, there would be more demos, and maybe a "proof of concept" pilot. Opinions were formed around the company about which pitch was best. Contracts were drafted, large teams were formed and a big kickoff session was held to launch the project. Some companies bought very expensive software this way – maybe yours did – some bought whole suites of applications that were ostensibly designed to work together as a company's entire ERP backbone. They then hired large consulting firms to implement the entire package, at great cost, which helped fuel the entire software boom of the last two decades.

Today there are many more choices, and more ways for vendors to market their products. Let's understand some of these choices and marketing strategies:

"Enterprise" packaged application vendors such as SAP, Infor, Oracle and JDA, function-specific vendors such as DemandTec (promotion management), Red Prairie (logistics) and SAS (data mining and business intelligence), and a host of other vendors make their living on the license-plus-implementation-plus maintenance-plus-future developments model, although many are also starting to adopt the Software as a Service (SaaS) model, which is a pay-as-you-go model, like a monthly or yearly subscription.

Packaged applications mostly fall into either the "horizontal" category (for example, a payroll system intended to work in many different industries) or the "vertical" category (intended for specific industries).

Large ERP systems generally combine both types of applications. For example, a company running an Oracle ERP system that wants to automate the sales function might choose to integrate the 'Accenture CAS' application for trade promotion management and sales analytics with its core Oracle software.

**You will find that many software vendors are trying to grow beyond their core expertise by claiming their solutions can also handle other functions within an enterprise.**

As an example, consider software firm Enterprise 21, whose advertising places it firmly in the category of manufacturing: "Discover Enterprise 21 Manufacturing Software Solutions." But its advertising also includes this claim: "Enterprise 21, however, is more than a stand-alone manufacturing software system - Enterprise 21 is a fully-integrated ERP system that encompasses order management, inventory management, procurement, RF and barcode-enabled warehouse management, advanced forecasting and planning, CRM, business intelligence, and e-Commerce functionality. All transactions and processes in manufacturing are directly linked to all other business departments and units throughout the enterprise with a single database to deliver vital, real-time business information."

What to do with this information, that this package can also replace systems for other functions? Some applications can, and some truly cannot. It's quite possible Enterprise 21's solution is a good fit for some companies for multiple functions.

My experience with these types of claims is that:

- While a package can indeed perform other functions, it is likely to do so with limited features because the ancillary features are not what the firm has spent years developing and improving, unlike the core features of the application.

- There are usually other vendors that specialize in those other functions.

- If you just want the core functionality that the application was originally designed for, you'll need to determine how to use just that portion of the solution while integrating it with the rest of your enterprise's software.

**Stick with a software vendor's competencies.** If you have defined your scope as one or two specific functional areas, look for applications that best fit those purposes. It's always interesting to look at a program's other features and functions, but unless you see hard returns in expanding beyond the package's main mission, stay within your scope.

**The biggest risk in overspending with packaged applications is during the implementation phase.** A single implementation of, say, a system to manage order processing, can cost $2 million to $3 million or more. Why? Two main reasons:

- You don't know exactly what modifications are needed to make the program work the way you want (see Case Study #6); and

- You don't know what delays you might encounter; each delay prolongs the project and adds billable hours to your cost from either the application vendor or other people you have hired to help with the project.

**Plan the life-cycle costs of a packaged application**. The life of a packaged application can span many years. The main costs will be annual support fees and any enhancement (custom development) work you might need as your business requirements change. Your vendor will also be releasing upgrades to the software, sometimes as frequently as every six or nine months, and you will need to stay current with those upgrades in order to get the best performing software and to continue your support contract. It's quite possible your vendor will suspend support on older versions, and that upgrading to the latest version may require you to install new hardware or a new operating system or both.

**A packaged application is not ready-to-use, in any sense of the word**. The word "packaged" is kind of a misnomer. You won't find a warehouse management system in a shrink-wrapped box on the shelf at Best Buy. A packaged application is simply one whose features and functions match *in general terms* what you want the software to do, but which still needs to be *configured* for your business. Configuration (also called setup) involves a lot of work.

Recall in Chapter 1 the dialogue between you and the software. All of the information needed by the software, in that example, is part of configuration. Much of the time involved in configuration is not in setting up all the data and parameters, but in deciding what data and parameters to set up. What data do you want to/have to set up for each customer? Should employees be "suppliers," so that you can reimburse them for travel expenses? Do you want to manage your inventory levels according to min/max parameters or days on hand, or some other way? In most enterprises, decisions like these aren't made by one person – groups of people get involved, and all those meetings and explanations have to be scheduled, and someone has to herd everyone into a decision. Not a quick process.

**Understand that software is an annuity business.** In simplistic terms a software company survives in the long run because it is able to collect annual maintenance and support fees from its customers while providing custom development services and a stream of version upgrades. The support fees are 20% to 25% of the original cost of the software license, so to the software firm, it's like selling a new system to the same customer every four or five years. In exchange for the fees, the customer gets access to support desks, can have the software firm make modifications to the system usually on a time-and-materials basis, and automatically gets some upgrades and patches (or fixes) as well as user guides and technical documentation.

Recurring support revenue is also highly profitable. In a recently released quarterly earnings report, Oracle reported that recurring maintenance and support fees accounted for 64% of its software revenue, but just 3% of its operating expenses.

Software companies (and the consulting firms that provide software services) also rent their people – people who have specialized programming or business process expertise, or expertise in servers, networks and other hardware. These people command six-figure salaries, and are usually billed out to a project at a rate of $150 to $250 an hour, plus travel expenses. One expert on a project for one year would cost you, the customer, roughly $400,000. That may seem excessive if the cost of that expert to the firm is, for example, $175,000, but it's rare that a firm can bill people out to a project for a solid 12 months. That expert will most likely not be able to bill, in a typical year, 40 hours a week for 52 weeks.

## Be Smart About Integration

Nearly all new applications you add to your enterprise will need some level of connectivity to your existing systems, and so integration is a key part of your implementation plan. The problem is this: every software firm will tell you their application can be integrated to just about anything. Yes, anything is possible, with enough money and effort. Here are some key areas you should be familiar with:

Application Programming Interface (API) - An API is a protocol used by software components to communicate with one another. It can be source code, or written specifications. A good question to ask the software vendor is whether it has developed APIs for interfacing with other programs, and which programs in particular. You'll still have work to do if the vendor has well-developed APIs, but at least you'll know someone sat down, thought about, designed, built and tested some form of integration – all work you shouldn't have to re-do. One area of the money pit avoided.

Electronic Data Interchange (EDI) is a standard for electronic messaging of commerce between two entities and frequently between two different systems. EDI was established in 1996 by the National Institute of Standards and Technology; it has widespread use around the world as a replacement for paper-based transactions between companies.

The application you're considering may use EDI as its standard method of interfacing other systems, and that is fine. But don't assume that EDI means it's as easy as plug and play, because while EDI is supposed to be a standard, it has been customized by enterprises for their own specific needs, and in addition there are actually two standards for every EDI message — UN/EDIFACT, and ASCx12. It's not unusual for companies to maintain a dozen or more partner-specific EDI data maps.

EDI, I believe, will eventually be replaced by newer technologies. EDI has limitations: It's prone to failure if data is incorrect or not recognized by the recipient's system, or if changes are made to its structure without thorough testing. EDI is also like sending data through a tube – no one sees the data except the sender and the recipient, so if you want to exchange data or transactions with another trading partner you have to build another tube.

**Don't underestimate the time and cost of integration via EDI.** My experience with EDI is that it takes a team of people to maintain it; to monitor the pipes, to push through transactions that have stalled or encountered errors, and to develop new connections or changes to existing ones. If EDI is how you will integrate applications, just be aware of the costs and effort involved. EDI is not standard, not necessarily low cost, and certainly not 100% reliable. If EDI is the way your customers want to exchange with you, you will have to accommodate. To mitigate the negatives of EDI, you can contract with one of the business-to-business (B2B) commerce companies that have emerged in recent years. These firms will host your EDI integration, monitor your connections 24x7, react to and solve messaging failures, and map new connections to trading partners.

**The quality of your master data will directly affect ease of integration across your enterprise**. Master data is data that should have a consistent definition and format across your organization. For example, a finished good or product in a manufacturing company should have the same spelling, Universal Product Code (UPC) and internal product number whether it is listed in a sales report or a manufacturing report. A gallon of Sherwin-Williams paint, for example, might have the following pieces of master data:

- Product Number:    0337719
- UPC                2780
- Description        'Sherwin-Williams Fine White Eggshell 1 gal.'
- Brand              SW Pre-Mixed
- Weight             10.4
- Unit of Measure    pounds

For applications to talk to one another, regardless of method, *they have to use the same definitions of data.* If the sales application uses "lbs." as part of the description, and the finance application uses "pounds," some type of translation will have to be made in the interface so that when the interface sees "lbs." in transactions coming over from the sales application, it converts it to "pounds" before it can be received.

If your master data is not standard across your enterprise, the extra code necessary to translate incongruous data will add to the complexity and cost of integration. These custom integration algorithms will be specific to your enterprise and therefore make your whole application structure more difficult for a new hire or a contractor to understand.

To further control costs and complexity of integration:

- Use a hub structure if multiple applications need the same data. This could be an "exchange" where pieces of data are posted then retrieved when needed by applications plugged into the exchange. This avoids creating spaghetti. The B2B services described earlier can serve as this type of exchange.

- Document everything clearly and insist that all documentation be kept current. Otherwise, you will end up with a complicated setup that only one person is familiar with.

- Let your ERP system define your master data standards, and keep those standards consistent in all legacy systems. If you have an ERP system, most likely it required you to define and set up master data. Since ERP systems usually span multiple functions, the master data setup is shared among functions and therefore common to departments like sales, finance and manufacturing. Keep these standards in place and use them to define the data transfer with any new application.

- Keep the number of integrations to a minimum and the integrations themselves as simple as possible! Frequently users will ask for pieces of data from other applications to be "integrated" to the application they most frequently use, and sometimes this is valid and has real value. But usually what the user wants is to simultaneously see different pieces of data – a job better managed by analytics software or simple query tools.

## Have A Structure For Evaluating Your Choices

An easy way to structure your decision making when it comes to selecting a software solution is to build a grid or a table, with vendors/solutions across the top and your most important criteria down the left hand side. You can weight how important the criteria are and develop a scoring formula. The result is an overall score that at least points you in the direction of the solution that best fits your needs.

The criteria will typically include cost and time to implement, but below are some additional checklist items I believe are very important.

**Table 1**

**A Software Solution Checklist**

| Checklist Item | Explanation |
| --- | --- |
| **Does it solve my problem?** | Does the software company's system solve your business problem? Does its existing functionality match the business requirements you drafted? |
| **Does it pay back?** | Do the financial benefits from the solution pay back the total cost of implementing it in three years or less? |
| **Do I understand all of the solution's costs?** | Have you accounted for initial license, recurring support fees, custom development costs for changes you want to make to the software, hardware costs, upgrades to your network bandwidth or operating systems on your current servers or PCs, the cost of the next version upgrade, the cost of consultants, of hiring backup staff for project team members, and travel? |
| **Is the solution in line with my strategy?** | Does the system match your criteria for what types of information solutions you will invest in, now and in the near future? |
| **Do I understand all of my alternatives, besides this particular vendor?** | Have you done your homework regarding software options available? Have you constructed an evaluation matrix and compared all the alternatives to one another? |

| Checklist Item | Explanation |
|---|---|
| **Does my team have the time and skills to implement this solution?** | Can you secure near full-time people to manage this project? Is the system easy to learn? Is it intuitive? Has your team evaluated it and are they comfortable they can master it? |
| **Do my users have the aptitude to learn it and become proficient?** | Can you envision your end users quickly learning to use all aspects of the software? Are there enough users who could become proficient enough to serve as key users and help other users with training and troubleshooting? |
| **Does my team fully understand how this solution will integrate with the company's other systems?** | Has the vendor demonstrated to your satisfaction the ease with which the system will integrate with your other systems? Are other enterprises already running the software with systems like yours? Try to get at least a conference call with those references to gauge the level of integration complexity. |
| **How risky is this particular software alternative compared to others?** | Can the software be phased in without interrupting the business? If the solution fails or the team encounters startup problems, how easy will it be to keep mission-critical activities running? |
| **Vendor reputation** | How many enterprises are using the vendor's software, and for how long? Get references and check them. |
| **Can I find programming help in the open market?** | If you need customizations, can you readily find people to do the work? Or are you locked in to using the vendor to make all your changes? |

# 9

# UNDERSTAND YOUR OPTIONS

"The most amazing achievement of the computer software industry is its continuing cancellation of the steady and staggering gains made by the computer hardware industry."

- Henry Petroski

In the 21st century, the software market is like the hordes of vendors on the streets of Marrakech. In Marrakech, shop vendors lure you with shouts, gestures, eye contact, price discounts and in-your-face displays of their goods. In the American software market, you are spammed with white papers, conferences, studies, forums, user groups, advisory studies and a legion of consultants whose fortunes depend directly on the proliferation of software products and a continued stream of investments by people like you.

And as I mentioned previously, the software market is rather incestuous; all players compete with one another for your business, but more importantly, they all hype the market in their own way. And the "players" are not just the software firms themselves, but all of the supporting consulting firms, industry advisory and research companies, and associated ad-supported magazines, newsletters, blogs and trade associations, all of whom hold conferences to bring people together for schmoozing, demos and eventually deals that bring more revenue into the industry.

It's just marketing, OK? I'm not trying to indict the industry; this is not a documentary exposing evil and corruption in the software market. I like these firms and the people who run them – they are creative, interesting and extremely sharp business-wise as well as technology-wise. Just understand that a lot of what you see and hear is part of how firms "go to market."

This chapter is a very brief overview of some of the software choices available in 2013. In no way is this overview intended to be a complete "buyer's guide" or anything like it. I present this section to provide a sense of what the market is like and how software products are geared for certain purposes and for certain market segments.

## ERP Software

An enterprise resource planning, or ERP, system is a collection of software applications that are all connected to each other so as to mimic and optimize the proper flow of data in harmony with an enterprise's business processes. The applications all share the same master data (customers, suppliers) and are usually already integrated with one another. An SAP ERP system, for example, has numerous modules, which are the function-specific applications inside the ERP, such as IM (inventory management), FI (financials – general ledger, accounts payable, etc.) and SD (sales).

In the beginning, when companies are small, they can manage with Excel spreadsheets. At some point, transactions become so numerous and pricing, customer information, formulas, vendor contracts and financial statements become so complex that spreadsheets, even with macros and pivot tables, become too unwieldy, inefficient and error-prone to use any longer.

So the enterprise implements software applications that manage the most labor- and data-intensive parts of the business. Soon it discovers that as good as these applications are, they aren't connected, which means the output of one application, which is needed by another application, has to be manually taken from the one and entered into the other, with the consequent loss of accuracy and timeliness.

An application that is used to keep track of sales might: capture sales orders from a sales rep or an online source; record the customer and delivery address information, as well as the method of payment; calculate the correct pricing and assign a delivery date; and calculate the net sales revenue to the company. It might then provide reports that can be downloaded into Excel, which are then used to order materials, send invoices, calculate profit, etc.

But this application by itself is an island; it has limited value, because it isn't connected to other applications the enterprise uses to order raw materials, calculate quarterly financial results, pay vendors and employees, and coordinate shipping. With an ERP system, these applications are connected. A forecasting system calculates expected demand, by customer, and sends this to another application that plans production and orders materials, and plans the shipment of goods from plants or vendors to warehouses. A financial application uses the purchase orders placed for raw materials to record accounts payable, and records the cost of product and shipping and overheads to calculate cost of goods sold, which it then records on the P&L as products are sold.

For a simple analogy, think about an auto assembly line. If the team that buys tires for the assembly line has just its own application for procuring tires, it is on an island. It has to receive information from outside the tire-buying software in order to use the application for its intended purpose. So the team might get Excel files in emails from upstream teams on the assembly line, detailing the production plans for each type of car, and when those cars are to be made. With an ERP system, the tire team sees that based on production plans, the tire-procurement software, which is now informed on an up-to-the-minute basis by all upstream plans on the production line, is recommending procurement quantities by tire type, by vendor, by day, and has even created purchase orders it can send electronically to tire suppliers.

**ERP projects are large and complex and therefore more prone to failure**. An ERP project involves a lot of users and requires a large number of business processes to be mapped to the new software, and consequently many billable hours by expensive consultants. These projects are launched with important names like Project Orion, Apollo or Zeus, big kickoff meetings are held, and the grand plans are laid before an enthusiastic team. I led such a project in 2002-2003. We had a 30-person team supplemented by another 20 people from the parent company; we rented an entire floor of an office building, traveled to Europe for training and planning, worked ridiculous hours and spent $13 million over 12 months, including $400,000 just on travel.

And that's about what it takes to implement an ERP system. Today, 10 years later, that type of project is still just as – if not more – expensive, but there are better and more efficient options for some companies in some situations ; and we cover those options in this book.

With all of that in mind, what follows is a summary of packaged application categories – certainly not comprehensive – that I have had experience with. I also list some packaged solutions in each category that are available today, sometimes with information from each vendor's web site. In no way is this section intended as an endorsement of any of the software products listed.

## Big ERP

Big-name companies Oracle and SAP offer traditional large-scope ERP solutions, which are entirely appropriate for big enterprises, global companies with multiple subsidiaries or single corporations with multiple plants and distribution channels. Typically these large ERP systems are hosted by companies at their own data centers, although third-party hosting is also common.

That SAP and Oracle are only good for big companies is no longer true, however. Both companies have made efforts for years to market to mid-sized companies with products that are simpler, easier and less costly to implement, and designed to compete in growing market segments. Oracle and SAP also continue to expand into hot horizontal areas such as business analytics, data virtualization, and in-memory computing (more on these subjects later). Oracle offers its JD Edwards ERP suite for mid-sized companies. JD Edwards was once an independent firm; its products have a solid history in the software market.

**Small to medium-sized companies that have the potential to become large firms through growth or acquisitions need the process, data discipline and scalability of an ERP system.** Just because your enterprise is, say, "only" a $200 million manufacturer with less than 150 employees, it doesn't mean you shouldn't consider the big ERP solutions. The advantages of big ERP also include an enormous user community with years of experience, readily available consultants, deep R&D and product development capabilities, and numerous choices of specialized applications on the market that have already successfully worked with Oracle and SAP.

## Other ERP

A lot of firms claim they have some form of ERP solution. But you need to look closely and make sure you aren't buying from a firm with expertise in one particular function, such as supply chain, that has decided to add some applications to broaden its reach in the market and – voila – repositions itself as an ERP solution.

Who are the top ERP providers? That depends. Do you mean client-server on-premises, or in the SaaS category, or ERP for small to medium-sized customers? SAP and Oracle are nearly always mentioned as the top two, and after that the list is less clear. Frequently mentioned are Infor, Microsoft, Epicor, JDA, Netsuite (100% cloud) and Metasystems. SaaS ERP, also sometimes referred to as "cloud" ERP, is not yet prevalent for large enterprises.

**Software vendors are either truly ERP providers, or specialized in one or more functions or industries with some limited capabilities in others**. Think of Home Depot as SAP: one place to go for all of your home improvement needs, with lots of breadth in product offerings and good quality as well. But say you need lighting – not just any lighting, but some unique fixtures to match the design of your home. In that case, you would probably go to retailers specializing in lighting.

Infor claims to be the #3 ERP software company with 70,000 customers (I've always wondered: does this mean 70,000 corporate entities, physical locations or end users?) whose ERP customers include Ferrari, Herman Miller and Brooks Sports. Infor has deep expertise in several key industries: aerospace, high tech, pharmaceutical, automotive, global retailers, financial services, and state and local government agencies. The company provides solutions as on-premise, cloud, or both.

From the company's web site: "From production management software for manufacturing, to comprehensive value stream mapping, our solutions provide all the necessary tools to hone your organization into a truly lean manufacturing operation. Our ERP solutions are designed by industry experts, many with decades of experience in their fields. Whether you produce highly complex products from distinct parts and components, or goods made by blending a variety of ingredients, Infor ERP software can help you reduce costs, improve operational efficiency, and give you the information you need to make better and faster decisions."

**Good reasons to consider an ERP system include**:

- You have spaghetti and it has a measurable and significant negative impact on the business. Spaghetti is a metaphor for numerous applications connected with one another by ad-hoc, non-standard means. Your application map looks like canisters connected to each other with spaghetti.

- You anticipate growth and your current applications, while still doing the job, are either home-grown or not supported outside your company. They may be applications that aren't available on the market anymore. All of your support is by people within your company.

- Throughout your organization, your people are spending an inordinate amount of time supplementing your systems with Excel and engaging in manual steps just to perform basic transactions. Your firm is growing fast and there are specific and measurable savings in administration you can easily identify. An ROI analysis is needed here.

It's hard to run a large enterprise without the end to end integration a good ERP system provides. Think of it as the main plumbing – every enterprise needs it. Specialized or industry-specific applications can then be added to provide functionality that the core ERP system doesn't have. But understand that it is a large undertaking. It will involve a lot of people in your organization for a year or more.

With an ERP vendor, it's even more important that the firm have a reputable product, a good support organization, a healthy and reference-able installed base, and a community of developers you can draw on for future needs.

## Manufacturing

These applications keep track of production formulas, in terms of quantities and cost, enable production scheduling, calculate and report on efficiency measurements like machine utilization and percentage of wasted materials, and in some cases measure cost of goods sold. These systems are usually linked to the core financial system. Subcategories include software to manage preventive maintenance schedules, spare parts inventory and ordering, and to manage and report on product defects and percentage compliance to quality standards, and labor management systems that track worker output and productivity.

One solution in this category is Oracle's JD Edwards Enterprise One Manufacturing Management. The company's web site claims the application "manages all manufacturing modes with a single enterprise-wide system where all manufacturing processes share common inventory, material, planning, purchasing, and financial databases."

Manufacturing execution systems (MES), a subset of applications in the manufacturing sector, are applications that directly monitor and control the manufacturing process; for example, managing the dosing of different materials into a vat of formula, opening valves, operating PLCs (programmable logic controllers), and displaying real-time graphics of the different stages of production. These systems usually also calculate and display management-related indicators, such as cycles per second, minute, hour or day, pieces produced per man-hour, and percentage of material losses in the production process.

Wonderware is an MES software brand with a long history in real-time monitoring and control of manufacturing processes. The company is owned by UK firm Invensys, and claims more than 500,000 licenses sold in over 100,000 manufacturing plants around the world.

The company's web site claims: "Wonderware is the market leader in real-time operations management software. Wonderware software delivers significant cost reductions associated with designing, building, deploying and maintaining secure and standardized applications for manufacturing and infrastructure operations. Our solutions enable companies to synchronize their production and industrial operations with business objectives, obtaining the speed and flexibility to attain sustained profitability."

Wonderware's customers include Chevron, Norfolk Railway, Nucor Steel, New Belgium Brewing Company, and Magna Automotive.

Another subset in manufacturing applications is shop floor management. These systems are used by contract manufacturers and companies that make to order specialty products. The software helps with estimating time and costs, managing schedules and resources, and coordinating material receipts and shipments.

E2 is a private company that claims to be "the authority on manufacturing software." The firm's web site states: "The E2 Shop System is comprehensive manufacturing software that puts total shop floor control at your fingertips. Designed just for job shops and make-to-order or contract manufacturers, E2 equips you to see your business like never before, and get the big picture on the best way to manage it."

Maintenance and maintenance parts are two other manufacturing sub-categories where the software market offers several competitive alternatives. This category is referred to as CMMS, or Computerized Maintenance Management Systems. These systems keep track of maintenance schedules for equipment, tell you when parts have to be ordered, manage vendor contacts and accounts, manage your parts inventory, and integrate to your financial systems to record parts expense and value of the parts on the balance sheet.

Like many software categories, parts and maintenance management has vendors that range from the big players to firms that offer free downloads and desktop versions. SAP, Oracle, and JD Edwards all have established offerings in this category. Here, my advice is to select something close to home; meaning a vendor who is either your ERP vendor or a well-known and established software company. The reason is that maintenance is usually a non-critical function, and doesn't warrant a lot of integration and customization work. The ROI for this type of application usually comes from saving administrative time in keeping track of maintenance schedules and keying purchase orders and receipts for parts into your financial or ERP system. Manufacturing applications are no different from other software in that they need data to deliver the benefits they offer. That means someone has to key this data in, unless the software is getting data from a connection to another application. Purchase orders, formulas, material usage, quantities produced, raw material receipts, and product master data all have to be entered into the system. This is one of the downfalls of software, not unique to manufacturing applications but maybe more prevalent in that category because the manufacturing floor is not the place you normally have the people who enter data: the quality, accuracy, and timeliness of the software's output is directly correlated to the quality, accuracy, and timeliness of the data going into the application.

## Supply Chain

Supply chain applications have been hot the past 10 years or so, following a trend in many industries to cut costs out of their distribution networks while reducing order lead time and inventories. Companies want solutions for modernizing what has traditionally been a backwater of many industries: truck booking; warehouse inventory management; pallet management; order picking; truck loading; yard management; and delivery discrepancy management.

This category also includes software for demand forecasting and optimal product deployment throughout a company's distribution network. As described in the Nike case study, demand planning software is fraught with peril for companies lacking the discipline and attention to detail needed to master these applications. And these applications can be difficult to evaluate, from a buyer's standpoint. Proceed with caution.

Companies in the supply chain space, known as SCM, for supply chain management, include software industry leaders such as Oracle and SAP, plus firms like JDA Software, Manhattan Associates and Red Prairie (recently merged with JDA). JDA is a firm that has grown by acquiring industry-leading supply chain management applications such as i2 and Manugistics. The company claims that 6,000 firms worldwide use its SCM software.

JDA's website states: "JDA's innovative supply chain management, merchandising and pricing excellence solutions empower customers to make optimal decisions that improve profitability and achieve real results. We're dedicated to providing customers with solutions that are easy to implement, maintain and upgrade, deliver innovation with a rapid return on investment, and have a powerful impact on their bottom lines."

JDA's customers include Kimberly-Clark, Ace Hardware, Fairchild Semiconductor, Macy's and Black & Decker.

You need to understand that supply chain applications can be single-purpose or inter-connected, like an ERP system. That's because the supply chain is like a business-within-a-business: it has at least five distinct processes – depending on your industry – that must all be interconnected in some way: 1) demand planning (what are we going to sell?); 2) distribution network planning (where should we send it before it is shipped to the customer?); 3) manufacturing scheduling (how much should we make, when?); 4) material requirements planning (MRP – what raw materials, supplies, or other materials do we need to make what we've planned to make?); and 5) warehousing and shipping (store the product and send it to the customer when needed).

A single-purpose application will claim to solve your problems in one or maybe two of these areas. An inter-connected supply chain application will manage all five of these areas and will inherently be inter-connected. Vendors that offer an inter-connected solution will present themselves as offering total SCM, or supply chain management, solutions.

It's hard to carve up the supply chain and say one application is better in one area than another, because probably the most important thing is the integration between the five main segments of demand planning, distribution planning, manufacturing scheduling, MRP, and warehousing and shipping. If you had to choose, the demand planning software might be most important, unless your customers place orders way in advance of shipment; on the other hand, you might have multiple manufacturing locations or warehouses and you need a solution for where to make and ship your product most efficiently.

## Finance

Finance and accounting functions were among the first to be automated through software. The sheer volume of numbers and calculations, reporting requirements, tax filings and payroll mechanics, plus the fact that nearly every business has to engage in these activities, made the area perfect for software.

When just these basic functions are needed, not much distinguishes one finance application from another. They all post transactions to a cost center and sub ledger account, they all capture sales and costs and calculate required P&L and balance sheet data, and they all provide reports. They might distinguish themselves in terms of ease of use or report writing, or banking account integration, or cash management, or some other aspect.

Many finance applications are simply bookkeeping systems; if you want real analysis you'll need to extract data to Excel, Business Objects, or another analysis and reporting tool. My own experience with both Oracle and SAP bears this out: even these leading finance packages are mostly concerned with accounting and financial, not management reporting. Oracle and SAP both have what they call "business intelligence" capabilities, but they are contained in separate modules that must be purchased and integrated with the core software. So companies can easily spend millions implementing SAP or Oracle, and still find themselves extracting data into Excel spreadsheets for basic business analysis.

My experience is that most finance applications lack budgeting and financial modeling capabilities. It is one thing to know that your prior month results were over budget because of rising fuel prices, and quite another to project the future profit impact of different oil price scenarios. At what point would it make sense to switch to alternative fuels, to pass on some of these increased costs, or to buy oil futures as a hedge? A typical finance application won't help you to answer these questions because they mostly record and categorize costs based on what already happened, not what might happen in the future.

Yes, there are "what if" modeling applications available on the market, but as a stand-alone application they aren't very useful, since you have to enter all of your data, as if you're using an Excel spreadsheet. The modeling application needs integration with your ERP to be most effective. Your ERP is the source of all kinds of data needed for financial modeling: production costs, formulas, material costs, transportation costs, revenue by product, as well as cost standards and budget information. This data changes frequently based on business conditions, competition, labor costs, and many other factors.

Microstrategy, Oracle Hyperion and Cognos are leading names in the financial modeling and analytics areas, but other, smaller firms are emerging. Netsuite, the ERP-in-the-cloud vendor, offers an add-on financial modeling application. Netsuite's web site states that the modeling application features these capabilities:

- Dynamic formulas and assumptions
- "Actuals" data incorporated into new forecasts
- Workflow management
- Planning of full financial statements
- Unlimited versions for "what-if" analysis
- Multi-dimensional models for complex sales and product planning
- Multiple currency budgeting
- Graphic drag-and-drop report builder
- Multi-version variance reporting (vs. budget, vs. plan, vs. forecast)

A3 Solutions is another, smaller firm offering financial modeling applications, either on-premise or as Software-as-a-Service.

A3 uses the Excel spreadsheet as the user interface, claiming it is the friendliest environment for creating what-if scenarios, and provides tools to link multiple sources of corporate data and manage modeling versions dynamically and virtually through its Spreadsheet Automation Server.

A3 claims McDonalds, Honda, Toyota, T. Rowe Price, and American Airlines as clients. Simplicity, speed of implementation, and low cost are A3's main selling points.

## CRM And Sales Management

Customer relationship management (CRM) and sales management are two other growing areas for software, with new applications frequently coming on the market as companies seek to know everything about their customers – what selling strategies work, which promotions produce the most sales lift, and how profitable their customers are. Companies have used CRM-ware in their call centers for years to instantly display a customer's account information, credit limits, preferences and order history. CRM is becoming more sophisticated, producing order pattern analysis, identifying distribution voids, and quantifying the impacts from competitors. Sales teams call on customers armed with more sophisticated data to support selling in a new line of products, perhaps showing the customer the product line's potential profits across the customer's 500 stores.

Consumer goods companies such as Kraft Foods and Procter and Gamble sell a lot of volume via specials and promotions at thousands of retail outlets across the country. It's just the nature of how demand is driven in the industry. Shoppers love bargains and retailers use deals to generate store traffic. A particular product, such as Kraft Macaroni & Cheese, might be on sale for a week or more, say, at Kroger stores in the eastern region of the U.S., at a price of three for $1.00. Companies such as Kraft might have *thousands* of promotions in effect for a given quarter of the year.

Think of the possible combinations of Campbell's soup (300 or more SKUs, 11 different brand categories), sold through most of the country's 36,000 supermarkets, on special at any given time, and the different possible deals, such as a temporary price reduction, a percentage off invoice, buy x get y, 2-fers, 3-fers, 5-fers, and so on. Who is going to keep track of all these deals? Software firms such as Siebel (once independent, now owned by Oracle) sell applications not only to manage these deals, account for the proper expenses and send the correct pricing to the invoicing system, but also to provide analytics to determine promotion effectiveness.

Other offerings in the CRM category include: demand estimating applications such as i2 and DemandTec (owned by IBM), whose sophisticated models predict demand lift from a given set of promotional activities; JDA's Vista, which manages the overall promotions budget and ensures promotion spending control; and SalesForce.com, the leading CRM-in-the-cloud application.

Banks and other financial institutions need a more business-to-consumer type of CRM and sales management. The top firm in this category, by revenues, is FIS Global, described on its web site as "the world's largest global provider dedicated to banking and payments technologies. With a long history deeply rooted in the financial services sector, FIS serves more than 14,000 institutions in over 100 countries. Headquartered in Jacksonville, Fla., FIS employs more than 32,000 people worldwide and holds leadership positions in payment processing and banking solutions, providing software, services and outsourcing of the technology that drives financial institutions."

## Analytics

Analytics is one of the top buzzwords in business software today. Analytics software is often marketed as a tool for business intelligence, data mining or insights. It's the proverbial crystal ball software: tell me things I don't already know, and show me ah-hahs or other exciting revelations that, if acted on, will increase sales, cut costs or produce some other benefit. SAS, based in North Carolina, has long pioneered this space, and now many business software firms claim to provide "robust analytics." The problem: what constitutes "analytics"? Canned reports are not analytics. So you'll need to shop this category knowing that probably the most serious applications will come from firms that are dedicated to analytics.

International Data Corporation (IDC) reports that the business analytics software market grew 14.1% in 2011 and is projected to grow at a 9.8% annual rate through 2016. IDC describes the market as dominated by giants Oracle, SAP and IBM, with SAS, Teradata, Informatica and Microstrategy rounding out the top 10 in terms of sales revenue. Although the top 10 account for 70% of the market, IDC reports that "there is a large and competitive market that represents the remaining 30%...hundreds of ISVs (Independent Software Vendors) worldwide operate in the 12 segments of the business analytics market...some provide a single tool or application, others offer software that spans multiple market segments."

Here are some other interesting analytics or business intelligence (BI) products:

Qliktech, which I mentioned in an earlier chapter, provides easy-to-develop dashboards with graphical representations as well as tabular and exportable reports. Its Qlikview software is an "in-memory" application, which means that it stores data from multiple sources in RAM, allowing the user to see multiple views of the data, filtered and sorted according to different criteria.

Information Builders (IB) is a software company classified by advisory firm Gartner as a leader in BI applications. IB's main application, WebFocus, is a flexible, user-friendly tool that is popular with sales teams because salespeople use it while visiting customers to enhance their selling messages with facts and visual interpretations of data. WebFocus has a "natural language" search capability, making it useful to monitor and analyze social media.

Birst, named by Gartner as a challenger in the BI space, is a cloud-based (SaaS) application that offers "self-service BI," deployment to mobile devices, adaptive connectors to many different types of data sources, in-memory analytics, drill-down capabilities, and data visualization. The Birst tool also has a data management layer, allowing users to link data, create relationships and indexes, and load data into a data store.

## Big Data

Industry players are currently hyping "big data," a term used to describe tons of data. We're talking millions or billions of rows here – way too much for standard query tools. What would constitute "tons" of data? Every bottle of "spring," "purified" or "mineral" water that was scanned at a grocery store checkout during the month of July 2011; the brand, the price, the size, the name and location of the store, and the day of the week it was bought. That's six pieces of data, multiplied by the estimated 3.3 billion bottles of water sold *monthly* in the United States. Another example: every credit card purchase of something that could be classified as "food" or "beverage" in the state of Florida over the last two years, by cardholder billing address zip code, retail outlet and day of the week.

The analysis of big data is made possible by two developments:

1. The continuation of Moore's law; that is, computer speed and memory have multiplied exponentially. This has enabled "in-memory" processing of huge amounts of data without retrieving that data from disk storage.

2. "Distributed" computing structures such as Hadoop have made it possible for the processing of large amounts of data to be done on multiple servers at once. It's the web search engine technology coming into the world of corporate software.

The hype may be justified. Big data does have potential and should not be ignored. With the right software, a virtual picture of the data can be painted with more detail than ever before. Think of it as a photograph, illustration or sketch – with every additional line of clarification or sharpening of detail, the picture comes more into focus.

Michael Malone, writing in The Wall Street Journal, says some really big things might be possible with big data: "It could mean capturing every step in the path of every shopper in a store over the course of a year, or monitoring every vital sign of a patient every second for the course of his illness….Big data offers measuring precision in science, business, medicine and almost every other sector never before possible. It could ultimately have an impact as great as mass production did more than a century ago—creating a new world of mass personalization of products and services. The big-data revolution is already happening, with hundreds of applications already in use, for instance, tracking millions of chickens from farms in Thailand to family tables around the world, or monitoring the location in real time of every emergency vehicle in a major city like Chicago. Over the next few years, it will spread across every industry and scientific discipline."

In their recently published book, "Big Data: A Revolution That Will Transform How We Live, Work, and Think," Viktor Mayer-Schonberger and Kenneth Cukier say that big data will provide a lot of information that can be used to establish *correlations*, not necessarily *precise cause and effect*. But that might be good enough to extract the value from big data. Three examples from their book:

- Walmart discovered a sales spike in Pop-Tarts if storms were in the forecast. The correlation was also true of flashlights, but selling more flashlights made sense; selling more Pop-Tarts didn't.

- Doctors in Canada now prevent fevers in premature infants because of a link between a period when the baby's vital signs are unusually stable, and, 24 hours later, a severe fever.

- Credit scores can be used to predict which people need to be reminded to take a prescription medicine.

Analysts at advisory firm Gartner think that big data is both 1) currently in the phase they call the "trough of disillusionment;" and 2) promising enough that its use in BI will grow sharply.

## Data Virtualization

Another emerging segment of the analytics software market is data virtualization (DV), referred to by some as Information-as-a-Service (IaaS), which enables access to multiple data sources, usually in real time, without the time and expense of traditional data warehousing and data extraction methods.

Forrester Research defines DV as solutions that "provide a virtualized data services layer that integrates data from heterogeneous data sources and content in real-time, near-real-time or batch as needed to support a wide range of applications and processes."

Traditional BI or analytics methods rely on some form of data warehousing, in which pieces of data are extracted, usually from transaction systems, transformed or "normalized" (i.e., "formatted"), and stored in tables according to some type of schema. "Customer Account Number," for example, may belong in the "Customer" table, and so on. As covered earlier in the book, building a data warehouse and getting it to work right can take years, and require substantial technical skills that even many midsized to large companies just don't have.

Over time I believe the traditional relational database technology will be replaced by distributed computing for analytics (like Hadoop). The underlying technology of distributed and in-memory analytics – object-oriented programming and the use of data objects accessed via "keys" – is orders of magnitude less costly than traditional database structures.

In simple terms, a DV tool is supposed to let you "see" sources of data in different applications and databases, and to "select" data from those sources for your queries or analysis. The advantage is supposed to be that you are not extracting data from their original sources, just "viewing" them inside the DV tool or layer to build your analysis.

While it's feasible to connect directly to external applications and other data sources, whoever owns or manages that application or data source may prevent you from connecting directly, for security reasons, or to avoid overloading the database, to avoid corrupting the data, or simply because the data is proprietary and the provider allows access only through an environment external to the data source.

Forrester sees an $8 billion market for DV software by 2014. Forrester notes that the current market is dominated by big companies such as SAP, Oracle, Informatica, Microsoft and Red Hat, and specialized firms like Composite Software, Denodo Technologies and Radiant Market.

## Enterprise Information Management

The term "enterprise information management" seems to capture just about everything you could think of in terms of in terms of data and/or information you could ever want, right? But this term applies to the indexing, searching and compilation of *information* (not necessarily data) from all of the places in your enterprise where documents might reside.

It's hard to tell where document/information management leaves off and analytics begins. This is part of the mashing up of software functionality that is going on in the market today.

Information is everywhere – in emails, presentations, documents stored on a company's server, individual user hard drives, servers in the cloud, etc. So traditional search software is completely ineffective, because it expects data or documents to be neatly organized inside a box where it can simply sort through data and return matches to your query.

Modern enterprise information management (EIM) software is different because it can search multiple and geographically and systematically separate sources according to terms defined by the user. It does this usually through a web browser.

The market for these tools arose because companies generated tons of documents without any "filing" standards, other than placing them on a corporate shared drive or on people's PC hard drives. As a result, it was almost impossible to assemble all documents within a company's four walls related to a particular customer, vendor, product, project, formula or activity. The ability to perform this type of search is especially important to legal professionals, who must respond to government inquiries or parties involved in litigation. This type of search is referred to as e-discovery.

Just a few years ago, I worked on a project like this, except it was referred to at that time as a records retention project, and we installed software from vendor L. The software was basically a search tool for the company's numerous internal file directories, and required the indexing of every file according to pre-established criteria, and the establishment of a document hierarchy and permission levels. It also assumed that all of the company's 1,200 employees would store all of their documents on the company's shared drives, and no longer use their PC hard drives to store files (this was not realistic).

Today, the company that used vendor L is implementing a different system that is capable of locating files anywhere within the company's network – shared drives, hard drives, emails. In less than 24 months, software that cost over $1 million to implement was rendered obsolete.

EIM usually includes e-discovery tools, and tools for managing content or knowledge, such as user guides, formulas, troubleshooting guides, business process steps or standard operating procedures, system diagrams, and documents critical to retaining official records.

Clearwell Systems, acquired in 2011 by Symantec, is a leader in the e-discovery field. Symantec also offers other EIM solutions. The Symantec web site says this about the Clearwell e-discovery application: "The Clearwell eDiscovery Platform, nominated as a Leader in Gartner's 2012 Magic Quadrant for eDiscovery, provides users with one seamless application to automate the legal hold process, collect data in a forensically sound manner, cull-down down data by up to 90%, and reduce review costs by up to 98% through the use of Transparent Predictive Coding."

# 10
# CONSIDER CLOUD, SAAS AND MOBILE

"If we have learned one thing from the history of invention and discovery, it is that, in the long run – and often in the short one – the most daring prophecies seem laughably conservative"

- Arthur C. Clarke, The Exploration of Space, 1954

Companies that stay out of the money pit keep up with what is changing in the technology market. If you don't force yourself to frequently look outside the walls of your enterprise, you'll reach for the same solutions you always have to solve new business problems, at the risk of missing some very promising alternatives.

## Cloud Computing

Cloud computing is being marketed as something new, but it's not. A cloud is simply a server – a computer you don't own or maintain – that sits somewhere other than in your building, that you access to run applications or store data. The loan calculators that are ubiquitous on the web run in a cloud that you access via the internet. Data backup services run in the cloud, as do other web tools such as Dropbox and Google Drive.

In fact, cloud computing, technically speaking, can be considered as any software you access that is functioning outside your desktop computer and outside any server that is physically on your company's premises or within your company's security firewall. SaaS Success Story #1 involved cloud computing, since the application functioned on a third-party owned and maintained server and was accessed via a secure internet connection.

If you read an ad that says: "Get more out of your data with business intelligence in the cloud," it could be a vendor selling a SaaS application, database hosting services or both.

**Companies with limited IT resources should always consider a cloud solution**. What is new in cloud applications is an expanded range of products and services – today some companies can run nearly their entire business in a cloud environment. Instead of investing millions in a traditional on-site suite of integrated applications for sales, accounting, logistics and human resources, companies can securely access these solutions as if they were residing on a server inside their building, but without the cost and maintenance of on-premise applications. Enterprises using cloud applications also do not need to employ a staff of experts to maintain, troubleshoot and periodically upgrade the server(s) or software – all of that is managed by the cloud application provider.

The cloud is made possible by high-speed internet connections and the huge decrease in the cost of computer memory over the past decade. Stronger security methods have also contributed to the growth of the cloud, although some companies are still hesitant to trust a third party with their sensitive corporate data.

**Inexpensive cloud applications are available if you can run your business with standard vanilla applications.** Enterprises can access what is called a multi-tenant version of the software – where several companies use the same instance, but whose transactions and data are separated from one another by functioning in a different location or "node" of the software. This scenario is also effectively software-as-a-service (SaaS), because you don't own the software, the vendor does, and you are simply renting it by paying a periodic access fee or a fee based on number of transactions or users.

Another option is for your enterprise to own its own customized software, but outsource the hosting of it on one or more servers and networks. This model has been around for years – many companies outsource the hosting of their applications to data centers. This is why cloud computing, for all its hype, is nothing new.

## Software-as-a-Service

Software-as-a-Service (SaaS) has been part of industry lexicon longer than cloud computing, and is hyped today about as much as the cloud, although I think has greater promise. Simply put, a SaaS application is one that you do not own, but that you use for your business. You pay for use of a SaaS application by month or year, by time spent using the software, by number of transactions conducted or amount of data stored, by number of users, or by a combination of these elements.

With so many options now available in SaaS form, it makes sense to explore these alternatives. Leaving aside the quality of the application and vendor for a moment, the benefits are: no server maintenance, no operating system license or maintenance, easy access via the internet, telephone support, custom development options, and near-seamless transitions to newer versions.

The downside: lack of control over things such as planned outages and a reliance on a third party for mission-critical applications.

Let's simplify this in plain-spoken terms: since advancements in software are occurring so rapidly, why wouldn't you want to "subscribe" to your applications, letting the software vendor continually update the application you are using? Without a large up-front investment to pay back, companies can more easily switch to other SaaS applications that arise in the market in the future.

Some forecasters predict that many enterprises will convert back to on-premise systems because of poor service from SaaS providers; that may end up being true but I don't think it would interrupt the long-term trend. There will always be a software provider somewhere that doesn't satisfy its customers. There is nothing inherent about the SaaS model that makes it prone to poor quality. If the vendor is a good vendor, it will provide good software, whatever the delivery model.

SaaS will continue to evolve and improve. Increasing bandwidth, improved security tools, and simple economics will make this happen.

A SaaS application still requires lengthy setup, fit analysis, possible modifications, testing and a carefully managed implementation. You should view a SaaS application as being no different from traditional on-premise software options when it comes to making it fit your business model. There are still up-front implementation costs and the same inherent risks as other types of software.

Two years ago, a company I am familiar with implemented a SaaS application designed to forecast demand lift from a range of promotional methods including pricing, advertising and special merchandising. The project took about 50% longer to implement due to inconsistencies in the model, access and performance issues, and reports taking way too long to run. These issues were resolved eventually, but the failures resulted in unanticipated costs, and the company was not able to use the application for its annual numbers-intensive sales planning cycle, the major purpose for which the software was acquired.

What SaaS brings to the software market will become, I think, what car leasing brought to the automobile market. In my mind – and in the minds of others – the concept of SaaS is logical, practical and long overdue. SaaS is merely the next step in the evolution of software.

**SaaS is challenging the paradigm that software is a thing you buy, take back to your office and install.** Looking back some day, we might shake our heads and wonder why any enterprise ever thought it had to *purchase and physically install a copy of millions of lines of code* that ran on a computer within its premises, just to transact day to day business. When we're done shaking our heads, we'll remember that the reason companies did this was because there weren't any other good options available.

The market is receptive to more and more SaaS solutions, and software firms are positioning themselves to offer those products. Most of the big, traditionally on-premise software providers now offer at least some of their applications in the cloud. For you, this will mean more choices.

**Are there any good reasons why you would want a traditional on-premise application?** Yes: security is one reason, in certain circumstances. A data center that your company manages on its own may not be a target for hackers, depending on the prominence of the enterprise, whereas large, well-known companies with massive server capacity have in fact been breached by hackers wanting to do harm to a large number of people. Even if your SaaS provider isn't one of these big targets, you – or more importantly your customers – may want the *relative or perceived assurance* of your own firewall surrounding your applications and your data.

Another reason could be that you already have economies of scale suited to on-premise hosting – plenty of server capacity, a built-in support staff, and maybe even developers on your team who are capable of building out the application the way you want it. If you are positioning the application to be used by several subsidiaries within your company, you may also want central on-premise hosting. Companies that do this usually want to tightly control modifications to the "core" system and also manage access permission levels among users, as well as the total number of users. These actions can reduce costs significantly.

But if your enterprise doesn't have a solid reason for on-premise hosting, a SaaS model should save you time and money, as long as it provides the features you need. Many companies just aren't big enough today to carry large IT organizations that manage servers....and to get the most value from your technology team anyway, you want most of your expenditures devoted to what are called "transform" purposes – developing new capabilities for the business – vs. "sustain" purposes, also known as keeping the lights on. Companies today that spend 80% of their IT budgets on sustain activities – keeping things running – find it hard to progress toward new technologies; they simply don't have the resources.

## Mobile Computing

See those app icons on your iPhone? Don't you think it makes sense to put corporate apps on an iPhone too? Yes, as long as the interface is easy and you don't have to type a lot on your tiny glass screen. Enterprise apps are going there, and that means the desktop/laptop is moving into the palm of your hand. This is a good thing, and it's not just about convenience. By mobile-izing corporate applications, the user interface *must* become easier to use, because you would never simply duplicate a Windows screen on your mobile phone or tablet, unless your users insist on clumsily tabbing and typing into each field just as they would if they were in the office. Putting enterprise applications into mobile device form will finally force software vendors to make their user interfaces simple, overcoming a big negative to using traditional windows-based applications such as SAP and Oracle.

Today, employees frustrated with corporate-issued Windows-based PCs are increasingly asking to be connected to their companies via iPhones, iPads and MacBooks. The ease of use of these devices stands in stark contrast to the buggy and slow machines corporations have used as their standard-issue PCs for decades.

Employees who are in the field, especially sales teams, also want some of the superior capabilities of Apple or Android products, such as robust design and video applications, to enhance their selling stories or engage on a different and more dynamic level with their customers.

Mobile devices, however, were not necessarily designed to connect to large enterprise systems, so if the promise of mobile is to be realized, new applications will have to emerge to enable this connectivity.

In the back office, 10- and 15-year-old ERP systems are still sitting in data centers, running on servers that require redundant backup, licensing costs for the latest operating system, and the time and attention of your IT team to keep them running and to manage upgrades to new versions.

Companies can now consider moving these systems to the cloud, or replacing them with SaaS applications. At the same time, however, they need to consider and provide for the demands of mobile access, which could point to a different set of applications altogether and result in a wave of "re-platforming" that would rival the move in the 1990s to ERP systems.

Mobile computing isn't just about convenience. An illustration:

A few years ago I was in a supermarket in a Boston suburb – the fourth one that day – to learn why the sales team for a global manufacturer wanted to automate its store visit routine. Deborah, the sales representative, showed me how she parked her laptop in the child seat of a grocery cart, and took it, along with a thick four-pound binder and a clunky handheld scanner, into the store.

Her routine was to make sure that every one of her company's 135 products sold to the retailer was on the shelf and placed correctly according to her company's "plan-o-gram," or shelf-set standard that it deemed was most appealing to the shopper. The plan-o-gram was contained in the pages of the four-pound binder.

Also, she would review the top 200 competitive products sold in that store for the most recent week, from a printed report, visit the back room to check her products' inventory levels, find the manager and convince him to fix any problems she saw, talk up the company's new products, and show the store manager how carrying more of her company's products or shelving them differently would generate more profits for the store.

To do all this, she had at her disposal:

- A 1990s-vintage bar code scanner that would tell her which items were missing once she had scanned all of the company's products on shelf at the store;

- The four-pound binder with plan-o-gram and other relevant papers, such as recent or upcoming ads or coupons, mailers, or new product descriptions, promotional schedules, etc.

- The printed reports showing the top 200 items sold for that store;

- Her laptop, which might have relevant documents or emails.

At stake in the 21st century for Deborah and her employer is *presence on the shelf,* which is where the rubber meets the road for any manufacturer selling its products through retailers.

Deborah was given 20th century technology to do a 21st century job. She needs data at her fingertips: how is this store performing compared to the company's forecasts? Which items are threatened by competition? When did this store last run a promotion and what sales boost did it create compared to what the company needs it to create to make the costs of the promotion worthwhile? Is the shelf kept full or is there an out-of-stock problem?

**You should care about the mobile trend because it could offer high ROI investments**. Car rental agencies today process returned vehicles on the spot, handing you a receipt almost before you can get out of the car. In the past, this process involved people typing away at a computer behind a counter and a bunch of back-office administration. The same is true with airline check-ins on your Smartphone.

The ROI for mobile will come from eliminating costs, or expanding your company's reach all the way to the consumer and in the process increasing sales and attracting more customers more often because you've eliminated barriers to purchasing your product or service. In Deborah's case, the ROI could come from being able to show the store manager the TV ad (of which he was unaware) that will break in his region for the July 4th weekend and who now will order 50% more to make sure his shelves and secondary displays are full.

Consulting firm Accenture has been working to perfect a technology that can determine, by scanning a photo of a retail shelf full of products, which products are missing. A simple Smartphone camera is used to take the photo. The idea is to lower the cost and time involved in retail shelf audits. Already, simple store audits can be done via crowd-sourcing services like Gigwalk, which recruits people for ad hoc work via smart phones and the internet. Gigwalk, whose pitch is "instantly mobilize people to do work anywhere," boasts clients such as BMW, eBay and Microsoft.

**Put in place some type of governance around requests for mobile access**. A company I am familiar with recently enabled its employees to access company email with their smart phones. The company is also testing a rollout of iPads to parts of its sales team. But the company has no mechanism yet for dealing with employee requests to access company systems from mobile devices. It is merely responding to requests on an ad hoc basis.

Many enterprises are likely addressing mobile in the same way. This approach, however, has the potential to incrementally

add cost and complexity to running applications without any mea-
surement of the benefits. Also, the passive approach means you
aren't thinking about the potential game-changing improvements
mobile computing could bring, beyond simple convenience to
smartphone users.

**A decision to invest in mobile applications should involve the
same rigorous process as any other software investment**: Step
firmly into the future, pick the right technology and manage ruthlessly (more on this in the next chapter). To keep risk low, choose
a small demonstration project to conduct. Look around at all the
apps being developed for mobile devices and see if any of them
spark your imagination in terms of applying them to a business
problem in your enterprise. How can you see your application
users benefiting from any of the Smartphone app technology?
Many print ads now display a bar code that can be read by smart
phones to obtain more information, provide a coupon or trial
offer, enter a drawing, or any number of other purposes that could
be imagined by marketers and advertisers – a seemingly low-cost
way to extend the value of the typical print ad.

**To capitalize on mobile, you need software on the device and
a way for the device to reach your enterprise systems**. The smartphone apps that let you check in for your flight are programs
made to run on your iOS or Android device, which then connect
over the digital network to an interface mechanism that is the
connection to the airline's main reservation system. Some companies have begun to create standard application programming
interfaces (APIs) for their enterprise systems to enable web access,
either by company employees, partners or the public. Companies
such as Mashery help firms set up these APIs to enable mobile
access.

The combination of these APIs, the internet and mobile computing platforms also enable new business-to-business connections
that could replace older technologies such as EDI and even XML.
Plus, the programs written for Smartphone, iPad and tablet PC

devices are often faster and easier to use than the typical graphical user interface that comes with most enterprise applications.

Mobile-izing your enterprise applications will usually involve purely custom software development unless your current software vendor has developed mobile applications for the software you're running. SAP, for example, offers several mobile applications, some that provide access to the core SAP enterprise application. SAP says its Retail Execution mobile app works with its enterprise customer relationship management software to provide "anywhere, anytime access to data from mobile devices...."

Oracle also has a number of mobile applications that complement its traditional enterprise software applications, such as Oracle Business Approval for Managers, a Smartphone app for managers to approve expenses, purchase orders and other pending transactions. Oracle also offers Smartphone apps that work with its CRM enterprise software.

**Use caution and investigate the full extent of mobile app functionality as well as system requirements**. A mobile app is likely to have only a fraction of the functionality of its sister enterprise application. That's OK as long as you understand that's what you're getting. Also, your enterprise version of the software may require add-ons or at least a web interface mechanism to share data with its mobile version.

Having said that, I think several of the mobile versions are worth examining, if only for the intuitive, graphical, easy-to-use smart phone interface, which you might consider expanding to a desktop version. I say this because the standard GUI interface for both Oracle and SAP, while functional, can be intimidating and confusing to the average user. With SAP in particular, users must learn the unique and structured way in which SAP transactions are conducted; often several screens or routines must be used to complete a single transaction.

# 11
# MANAGE
# RUTHLESSLY

"There is nothing so useless as doing efficiently that which should not be done at all."

- Peter F. Drucker

The software money pit has a gravitational pull, and staying out of it is like war – against those factors that can delay, disrupt or kill your project. The fact that you have a *good* project team isn't sufficient to give you a win. You need experienced, strong and ruthless people on that team.

By ruthless I don't mean cruel or stubborn; I mean *an unwavering opposition to things that get in the way of success.* An unfocused team gets in the way of success. An unclear or constantly changing scope gets in the way of success. Indecision gets in the way. Managing ruthlessly also means vigilance in looking for emerging threats to your success.

By now you should be well aware that messing with new software is like playing with fire. Many organizations underestimated the problematic nature of software – ever since its birth as an industry in the 1950s – and have been burned. In the years since the 1950s, many good books have been written about managing software projects. Some are very comprehensive, explaining all of the details of software project management. These books are great for IT project managers and consultants who do projects for a living.

But for you, a business leader, I want to lay out in this chapter how to smartly manage even big IT investments without being an expert IT project manager. The key is a combination of your outlook, attitude and engagement with the project team.

- Your outlook is big picture; you focus on the few things that drive success and you stay out of the weeds.

- Your attitude is that this is a high-risk endeavor best managed by professionals; it can damage the business if not done right, and you are determined to get the ROI or the project isn't finished.

- Your engagement with the team is firm but supportive; you insist on structured work, discipline and facts, while doing what is needed to minimize distractions, supply the team with needed resources and eliminate barriers to their success.

## Create A Ruthless Team

The strength of your team is by far the most important factor in creating a winning investment. If you cannot form a strong team to carry out an important and expensive software investment, you should not make the investment.

**Your project manager (PM) is someone who is expert on the business processes you are trying to change,** and preferably also familiar with the systems currently used to enable those processes. He or she understands your business and how everything comes together to make it work. Your PM is a leader, is energetic and positive, and is skilled at assimilating many details into a single understandable picture.

**Don't reach for the logical person based on the organization chart**. Do not designate someone as project manager because he or she is nominally in charge of something closely related to what you are trying to do. You need, above all, 1) knowledge of the business and 2) leadership. I understand that in many companies this combination does not exist. If so you may have to source a project manager externally. I have seen situations in which PMs are selected who have neither the skills nor the aptitude to manage the project. That's not OK. You are spending serious money. Don't do it.

**Your PM needs two experienced and competent partners**. If your firm is like most, it's not every day that you make software investments involving big projects lasting months or years. Most everyone in your enterprise – business and IT – is busy running the business. No matter whom you select to manage your project team, this endeavor will be different, challenging and a big change from everyday work.

The two partners your PM will need are 1) the most senior and experienced IT person in your enterprise whose functional (business) expertise matches the nature of the project; and 2) an experienced implementation professional, preferably from a leading technology services firm, such as IBM, Accenture, Deloitte, Booz, Cap Gemini or Computer Sciences (CSC), although there are many good firms and many good independent consultants. **Give your PM the benefit of a partner who has done this before, preferably many times before!**

**Like you, your PM must be ruthless** – again, not mean, abusive or unreasonable, but ruthless in his or her intolerance for the usual project killers and for anything else that threatens success. Think about an attack dog. Attack dogs are trained to sniff out threats and obey commands to attack a threat. Your PM must have the instincts of an attack dog and the reasoned mind of a human to recognize which threats are real and which can be ignored or managed. You cannot manage ruthlessly if your PM isn't ruthless. Installing a weak or inexperienced leader as your project lead is one of the biggest mistakes you can make.

Your PM is a person who attacks the one-day delays as aggressively as the three-week delays. Your PM must also be OK with criticism or anger as a result of his or her fastidious attachment to the due dates on the calendar.

Your PM's ruthlessness must also extend to other areas: resolving issues that threaten success; securing resources (people, money, equipment, and services) that the project needs; and setting firm expectations of deliverables by team members and others.

Your PM must have the moxie to remove or go around obstacles, and attract and manage the attention of people in the room. With attention deficit disorder rampant in the workplace, not every project lead will be comfortable demanding the attention of others at the risk of being mocked as overbearing or unduly critical of people.

If by now you think that the ideal project manager as described is a rare individual, then you also get the point that most enterprises today are not prepared to invest in big software projects because the people they have available to implement said project are not up to the task.

**Let your (strong) PM choose the team, but oversee and adjust as needed**. If your PM is strong, he or she will know which people in the organization he or she needs for success. Allow the PM to state his or her needs in an unconstrained way, but get an estimated percentage of time commitment needed from each team member.

**Your job is to ensure your (strong) PM receives the needed resources**. You must use your position or influence so that other functions in the enterprise will make the needed commitments to the project. Make sure the right business functions have representation on the team and that they share equally in the accountability for success. Look at the team as a whole. Are they ruthless, or do they at least have the ability, when necessary, to be ruthless?

**Give the team the mandate in the strongest possible terms**. The project team must know and feel that they are on a mission that is critical to your enterprise. Keep the mandate simple: "Your mission is to move these five business processes to this application, and in the process fully realize these three main business benefits." *Nothing motivates a team more than a crystal clear goal stated in person by a very senior individual – preferably the top person in the enterprise.*

**If your team is always telling you everything is going well, something is wrong.** It's a fact: everyone wants to look good. Marcia Martin, a top executive coach and expert in transformational leadership, says "your mind, as a human being, wants to be right; in real everyday life it's as natural as breathing that we want to be right, look good and know the answer."

No wonder project leads, as human beings, tend to paint a more positive picture of things when managing the communications upward. But this is dangerous for system projects – real issues can be glossed over, delays can be explained away. You must be able to see through, around and over this. I can tell you that as a PM myself I kept a lot of issues away from upper management view, because I didn't want them micromanaging, and I didn't want them to hit the panic button about issues or delays I knew we could manage and resolve on our own.

**Don't let your team get lulled into the schedule preferred by your consulting or software partners.** This type of schedule is methodical and lengthy. Methodical is good, but lengthy is unnecessary. Your team should attack and complete the tasks in front of them as quickly as they know how.

## Embrace Structure

Think of the typical meeting in your enterprise: people come in late, sometimes talking on a phone or otherwise distracted, and 10 or 15 minutes are spent getting into the real content of the meeting. Maybe someone asks what the meeting is about. The person who set up the meeting speaks up and may (or may not) explain what is needed from the meeting. Discussions ensue – some are on point to the reason for the meeting; others are irrelevant. Agreement is reached on a few actions and follow-ups, some people have to leave early, and the meeting breaks up.

These corporate-casual norms are anathema to a serious and successful software investment project, which demands rigor and, above all, structure. Structure in terms of:

- A work plan with tasks, due dates and responsible owners that is reviewed daily;

- System requirements, testing scenarios and future processes that are documented in detail and agreed to by all stakeholders;

- An issue resolution process to ensure that impediments to progress are eliminated;

- Commitments by team members to own and deliver key parts of the project on time;

- An overall sense of urgency to complete tasks and kill barriers to progress.

A sense of urgency and focus are hugely important to success. The film *Apollo 13* – a drama about the safe return of a NASA spacecraft and its crew after the craft was damaged and the moon landing mission aborted – captures these attributes perfectly. While the whole film is a study in focused, man-plus-technology effort and achievement, one scene in particular drives home the importance of urgency and focus.

In the scene, about six technicians in a conference room must assemble an air scrubber exclusively from parts that are available on board the returning spacecraft, because the main scrubber in the craft has failed, causing carbon dioxide inside the craft's cabin to reach dangerous levels. The astronauts will die – soon – from lack of oxygen if they cannot fix or replace the scrubber.

The technicians jump on the challenge, and produce a prototype that the Apollo 13 astronauts are able to replicate. We witness urgency and focus, and a breakthrough that saves the crew.

Do you think any of the technicians in that scene would have said, "Hey guys, I'm good for another 10 minutes but then I've got another meeting"? No, the average software project is not a moon mission, but it does require the same kind of focused and structured work that such a mission benefits from.

About 10 years ago I was involved in a large ERP project. The Europe-based headquarters PM had temporarily relocated to the U.S. to direct the North American part of the project, which was on a 12-month timetable. Only three weeks into the project, everyone was in the "self-training" phase, which involved exploration of the software with test data and immersion in documentation of the software's functionality.

The self-training phase would yield a comprehensive set of gaps – areas where the software did not meet the needs of the business, and thus areas that would need eventually to be resolved somehow. In one of the early status meetings, I watched the Canadian project manager ask for a few more days beyond schedule to complete this phase, and was surprised to hear the headquarters PM flatly and vehemently reject the request. An argument ensued, but the HQ director prevailed. I understand now why he was adamant. Allow a little slippage and control is lost. The tone was set. As software legend Frederick P. Brooks wrote: "How does a software project come in six months late? One day at a time."

**You need a structured methodology, but not necessarily the one your software vendor or consulting partner uses.** A methodology is simply a way of doing things. A methodology for doing laundry could be: 1) separate colors from whites; 2) wash whites in hot water/cold rinse with bleach; 3) wash colors in cold water; 4) hang dry delicate fabrics and put everything else in the dryer on medium heat for 30 minutes. But there can be variations in laundry methodology depending on preferences and beliefs, such as 1) throw colors and whites together in a cold wash and cold rinse so colors don't run; 2) throw everything in the dryer on delicate cycle; 3) check periodically for dryness; and 4) pull out anything that seems dry.

IT project methodologies are practically an industry – a plethora of templates, software, books, training programs and consulting engagements. Which methodology is best? For your project, the best methodology is the one that provides structure in a common-sense way with simple, easy-to-use tools. A methodology encompasses both the sequential steps to a project, and the documents used (tools) to manage the different events in the life of a project.

The chart below explains the steps of a typical IT project methodology, along with a description and the tools used for each.

**These steps might seem geeky and overly serious, but it is within these phases that many projects fail**. See in the far right column how, in each step, you can be drawn into the money pit, and how to dodge the pit and win instead.

**Table 2**

**Basic IT Project Methodology**

| Methodology Steps | In Layman's Terms | Tools Used | How You Can Lose/How To Win |
|---|---|---|---|
| Current and Future Process Definition | How things work today and how you see them working after you implement the software. **Do this before you commit to a system!** | Process flow diagrams created in flowcharting software, PowerPoint or on flipcharts. | Often this is not done in enough detail, or is not done at all. Insist that it be done with great specificity. |

| Methodology Steps | In Layman's Terms | Tools Used | How You Can Lose/How To Win |
|---|---|---|---|
| Business Requirements Definition | Exactly what you want the software to do, in each specific business scenario. **Do this before you commit to a system!** | Excel document with columns for business process, functional requirement, explanation, and priority level. | Often not complete, or allowed to grow as the project advances. Freeze the requirements at some point. Recognize and stop wish lists. |
| Gap Analysis | Comparing what you want the software to do with what it can actually do, and documenting the 'gaps.' These include integrations to other systems. **Do this before you commit to a system!** | The same Excel document used for Business Requirements, with additional columns for "Fit – Yes/No," "Resolution" and "Workaround." | Failure to fully investigate the software's capabilities leads to overly optimistic fit evaluations. Winners do this step before they buy the software. |
| Gap Resolution, or "Gap Killing" | Deciding what to do about every instance where the software doesn't do what you want it to do, which means either you live with the gap, work around it somehow, or spend the money to modify the software to your exact needs. **Determine the cost of modifications before you commit to a system!** | This is completed in the "Resolution" and "Workaround" columns of the Gap Analysis. | Balance is needed here. Modifications cost money. Question your process first. High-priority developments should be for customer and total company reasons, not convenience or ease of use. |

| Methodology Steps | In Layman's Terms | Tools Used | How You Can Lose/How To Win |
|---|---|---|---|
| Development or "Build" | Program the changes into the software that will close the functionality gaps. | A very specific technical design document is needed here, usually Word and Excel. | The usual pitfall is that this is not specific enough or that specs do not accurately reflect the business requirement. To win, limit the modifications and don't make the software do something it was not designed for. |
| Configure Test Environment | Put a test copy of the software, including the changes you've made, on a server and set it up as if you were going to use it in your business. | A simple Excel list of all the data fields required by the new system and what you will put in those fields. | Can take way longer than expected, because people haven't made decisions on what data to put in each field. Winners hammer out master data, parameters and settings during the gap analysis or earlier. |

| Methodology Steps | In Layman's Terms | Tools Used | How You Can Lose/How To Win |
|---|---|---|---|
| Unit Testing | Mostly a test of the basic functions of the software to make sure all the setups and integrations are working. | A tracking spreadsheet. | Modifications from developers that still have technical glitches. Condition developer payment on quality and timeliness of work. Insist that developers work in your time zone. |
| User Acceptance Testing | Full testing of the software by people who will use it, performing transactions just like they would do in their everyday work. | A list in Excel of all the testing scenarios with corresponding columns for test results (pass/fail), dates, user name, transaction document number. | Incomplete list of test scenarios = surprises later on that the software doesn't do what you expect. Have a wide range of users sign off on the scenarios to be tested. Don't forget the exceptions – deleting a sales order, incorrect master data, etc. |

| Methodology Steps | In Layman's Terms | Tools Used | How You Can Lose/How To Win |
|---|---|---|---|
| Training | Complete hands-on training for everyone who will use the new system. | Complete standard operating procedures and step-wise instructions on all transactions, plus quick reference guides. | People are casual about the training, but after Go Live need lots of support. The only truly effective training is on the job in real-world conditions. Insist that a group of users participate in the testing phase. |
| Go Live Prep and Business Continuity Plan (BCP) | This is defined in a detailed hour-by-hour illustration of events that describe shifting the business from the old system to the new system. Includes uploading master data and all pending transactions into the new system; also includes plans to recover from failure. | A punch list in Excel of all tasks needed for go live prep, plus a depiction (PowerPoint or other) of each day or day part showing the sequence of required events. | The danger here is that this work is not done with enough forethought or detail – or isn't done at all. Business failures during Go Live add to project costs because of inadequate planning for system backups. Make sure you can still run the business even if the whole thing falls apart. |

| Methodology Steps | In Layman's Terms | Tools Used | How You Can Lose/How To Win |
|---|---|---|---|
| Go Live | The cutover from the old system to the new system. Usually on a weekend or holiday when the business doesn't need the system. | A record of the number of planned transactions and the number of actual successful transactions. An Excel log of issues or problems that arise with planned actions to resolve. | Not enough support people in two areas: 1) system experts to help users; and 2) backup employees for those who are distracted or slowed down by learning the new system. Staff up for several weeks. |

If you just scanned over the table above, go back and absorb it. *Understand the flow of events in a software project.* Use the table to compare against the project you are in the middle of or the project you are thinking about. If you need a more condensed version to commit to memory, think of it like this:

**Before you commit:**

- What does the business need and what does the software provide? Document the differences, and the resolutions to those differences – fix with customization, or live with them?

- With the best option out there, what is still missing and how will I fix that? Does it solve my problems and give me the ROI I expect?

**After project launch and for the duration of the project:**

- Does the software test out the way we expected? Do users understand how to use it? If the answer is no, you have more work to do – determine why and fix it; otherwise you aren't ready for Go Live.

**Before the switch over to the new system, ask:**

- Have we planned in detail for the Go Live, including what we'll do if things go wrong? Same thing applies here – if the answer is no, or even if you sense people are glossing over problems or making excuses for not being ready, you really are not ready and therefore risk the business disruption that can turn into a disaster.

**Managing ruthlessly doesn't mean imposing impossible goals or hammering the team with a constant barrage of questions and expectations. In fact, don't burden the team with excessive administration**. Project administration is important – every project needs a traffic manager with good tracking tools who is in control of the project plan and the status of each event in the plan. But don't go crazy documenting everything in sight. Typically Microsoft Project is used for IT projects, so I'll commit heresy here and say MS Project is not the best tool. It is maintenance heavy and OK as long as you have a full-time person to manage it, and you believe in the basic "waterfall" method of IT project planning.

The waterfall methodology used in MS Project and adhered to for decades by many in the project management field is arguably flawed from the beginning, at least according to some. This methodology assumes only that task completion is a function of the number of man-hours, total time duration, and dependency on completion of one or more preceding tasks. Viewed on a chart, the timeline of the project looks like waterfalls cascading from left to right, top to bottom, as time progresses.

Frederick Brooks asserts in *The Mythical Man-Month* that "the basic fallacy of the waterfall model is that it assumes a project goes through the process once, that the architecture is excellent and easy to use, the implementation design is sound, and the realization is fixable as testing proceeds (but) experience and ideas from each downstream part of the construction process must leap upstream, sometimes more than one stage, and affect the upstream activity."

Brooks also states the planning process usually and mistakenly assumes that the people needed will be there, available, focused and engaged, when they are needed, and that when they perform their work they will do so in a linear fashion, with no mistakes and no need to backtrack and start over.

The $13 million SAP implementation I managed 10 years ago was run with Excel, Word and PowerPoint. Excel has plenty of flexibility to track every detail, is easy to use and update, and can be shared with everyone.

Also, if at all possible, please avoid creating a "steering committee." I loathe steering committees, because:

- They tend to second-guess the team and impose their own actions and suggestions;

- The team feels obligated to create an impactful PowerPoint presentation for each steering committee session;

- Posturing occurs at such meetings, both by the project team that wants to look good and by members of the steering committee, who want to be seen as contributing wise and valuable input;

- Such posturing disguises real and serious issues that may threaten success.

If you or someone such as the CEO insists on a steering committee, keep it as small as possible and the updates as simple as possible, and encourage honest, open and non-threatening dialogue.

## Make People Pay Attention

In the 21st century, few companies are willing to dedicate large teams of full-time employees to a large system implementation; they just can't afford it and don't see it as a priority compared to day-to-day business. Since 1996, I have worked on many projects and it seems with every new project, the people resources available are fewer and fewer.

**The critical factor now, especially in the virtual age, is not how many people you can get on your team, but how much of their focused attention you can get when you need it**. As columnist and Guinness record holder for Highest IQ Marilyn vos Savant said, "Working in an office with an array of electronic devices is like trying to get something done at home with half a dozen small children around. The calls for attention are constant."

It is not easy to get people to pay attention. Your project is competing for share of mind with texts, email and the internet emanating non-stop from your team's cell phones. Thomas Davenport and John Beck, in their book *The Attention Economy*, state: "The ability to prioritize information, to focus and reflect on it, and to exclude extraneous data will be at least as important as acquiring it." Think about those words: *prioritize, focus and reflect, and exclude extraneous data*. Is that not difficult in today's business environment, characterized by a flood of information?

You might say, "Of course people can focus their attention!" A story:

In 2005, I was on a team whose mission it was to modify parts of SAP to enable the sale and invoicing of what are called "kits." A kit in software parlance is a product whose components include other products. The final product is what is recorded as the sale and what appears on the customer invoice. Sounds simple, right?

A team was needed to resolve questions such as: How will the assembly transaction and cost be treated – as a manufacturing cost or as a service paid to a vendor? Should we record as sales the component products used in the final product? Developers would have to incorporate these changes to the software.

Our company was global. The team participants were in Ohio, New York, Western Europe, Manila and Texas. After *14 months* the team had finally gotten to the point at which it could test the solution. There had been more than 25 full-team conference calls, but no face-to-face meetings in which everyone – business people, developers, and technical experts – was present.

Why did it take so long? First, for each team member, this small project was one of many things on their plates, which resulted in attention deficit. Second, the project was managed primarily by conference calls. The effectiveness of conference calls, I believe, is about one-third that of face-to-face meetings where software projects are concerned.

Do you know anyone who does not multi-task during a conference call? The attention deficit is automatic. People are caught not listening. Participants tend to speak a lot less than if the meeting is held face to face. It's easy to hide and withdraw on a conference call.

Your PM, being ruthless, is the detail-obsessed killjoy in the meeting, the prude, the taskmaster, or whatever is needed to whip your team into a serious, attentive and responsive state.

## Kill The Project Killers

Project killers are normally not external events outside the control of your enterprise. *Project killers stem mostly from the decisions made by management or by the project team:* the decision to attempt a complex solution, the decision to select software before thoroughly investigating its capabilities and fit, the decision to let business requirements grow unconstrained.

Failure often stems from the original plan itself – unrealistic timelines push teams to cut corners; the solution, the team, or the business isn't ready; management forges ahead and things fall apart. As mentioned earlier, author David Yardley describes in his book the failed London Ambulance Service system upgrade. While many factors contributed to the collapse of the project, Yardley first cites schedule pressure (unrealistically short implementation time frame) as a major contributing factor.

Yardley advises: "There is never a perfect time to implement a new system, but too many systems are planned and implemented at the very worst time – when the existing system is creaking under the strain of increased workloads and there is not enough time to plan a replacement system. The LAS computer-aided dispatch project fell into this category. The phased implementation that was introduced did not come about through planning, but as a result of desperation."

The table below lists the most prevalent project killers and how to deal with them.

**Table 3**

**Project Killers**

| Killer | How to Stop, Reverse, or Go Around It |
|---|---|
| Wrong technology selected, or the software needs way more customization than you thought | If it's too late, do what you can with what you have, no more. Don't try to make something fit that doesn't belong there. A narrow, limited implementation is better than a force-fitted overly modified solution. |
| Unrealistic deadlines for implementation | Narrow the project scope to what is absolutely essential. Deliver the rest of the project in later phases. Request more time and document the risks to the enterprise of not being ready on the planned cutover date. |
| Overly complex solution design | Stop and redesign the solution. Evaluate whether the complexity is really bringing sufficient value. Break the project into manageable parts. |
| Inexperienced project team or lack of knowledgeable people for the project | Stop. Get the people you need or don't do the project. You're just setting yourself up for failure. |
| Slipping deadlines leading to a late project | Triage. Concentrate on those parts of the software that are essential; leave the rest for later phases. Lock the team in a room and stop the interruptions. Clear away barriers. |

| Killer | How to Stop, Reverse, or Go Around It |
|---|---|
| Lack of user involvement | This can kill a project two ways: 1) solution design is flawed because the right people weren't involved; and 2) users don't know how to use the system. Designate two or more (as many as you can get!) key users who become experts *before* Go Live. If it's too late and you've already cut over to the new system, bring in as much one on one user support as you can. |
| System gaps not known before Go Live | If you've already cut over to the new system, you have no choice but to find workarounds, which are just manual substitutions for work the software would do. Staff up to keep the business going; get the modifications you need ASAP. |

## Learn From Your Own Experience

"It's all about throughput," I said to the facility manager. I was concerned about his crew's ability to keep shipments flowing once we converted the warehouse over to the new radio frequency technology and related software. We were in a high-volume warehouse in eastern Pennsylvania that shipped to very demanding retail customers up and down the East Coast and into New England. These customers could send your fully loaded 18-wheeler back the 350 miles it had just traveled, to your facility, because....you were late.

My fear about throughput stemmed from a thorough drubbing my team had gotten just a few months earlier when our software cutover on another project brought manufacturing lines to a halt (see Case Study #1). Yes, it was all about throughput. Keep the warehouse's activities flowing at normal speed and we would be OK. I clung to this maxim, emphasized it daily, and I think it was one of the reasons for a smooth Go Live in the Pennsylvania warehouse.

Because your enterprise will do this again – invest in software – a post-mortem should be conducted after the dust has settled and people have had time to reflect on what went right and what went wrong and what they would do differently next time. This helps you to avoid stepping in the same doo-doo.

**Did you really get the ROI you planned?** Not what you would present to your boss' boss, but what did you *really* get, compared to what your business case – that document you made up to get the money for the project – said you would get? This is a little harder than a post-mortem, and you may need six months or a year to get an accurate picture. But a clear-eyed acknowledgment of both the gains you got and fantasies that didn't come true will help you (next time) to avoid making one of the classic IT investing mistakes – fooling yourself about the ROI.

**What do the users think about the project?** At the risk of unleashing a firestorm, you need to get feedback from the application's users. You'll know in advance what the general tone will be, so tailor your questions accordingly. Remember, your enterprise didn't undertake this project to elicit joy from users – although they should see some of the same benefits you anticipated for the project. At first, users may simply hate the new application (this has happened on several of my projects). Large ERP systems tend to be a turnoff for users, at least initially, because of their demanding accuracy and structure. Separate complaints from useful feedback and file them away for future use.

**How did your software vendor(s) and implementation partners perform?** Would you use the same people and firms? Take out the contract and see if everyone did what they said they would do. Review it with your partners to see if they agree with your assessment.

# 12
# SURF THE
# TSUNAMI

"We are continually faced by great opportunities brilliantly disguised as insoluble problems."

- Lee Iacocca

Harvard researchers have successfully encoded a 53,000-page book into molecules of DNA. A *billion* copies of this book – in a liquid or salt form – could be stored for hundreds of years inside a small test tube. According to the project's lead researcher, the entire internet could be stored on a device the size of your thumb.

An electronics company has developed a way to laser-apply an RFID chip with a small antenna to various objects. Airbus is expanding its use of RFID part-marking to all of its aircraft, starting with 120,000 life vests and 40,000 seats in 2013. The RFID label or chip will immediately identify the presence and location of the part, and relevant data such as maintenance history or expiration date. A major Japanese clothing retailer has applied RFID to everything in all its stores, including its inventory, hangers and merchandising displays.

Via webcams, software can now trace the eye movements of a shopper as well as the shopper's path throughout the store, recording each step of every shopper to create a "heat map" showing what parts of which shelves shoppers are drawn to.

Yes, there are some very cool things happening in technology and software. But what does this mean for the typical enterprise? Before we answer that question, do YOU even know – whatever your function is in an organization – *what software technology is being developed specific to your field* that you could use to your advantage?

**To ignore software developments in your field is to be left behind professionally.**

## Pay Attention To The World Around You

Today it's not enough to be a good accounting or finance professional, a senior manufacturing manager or sales manager, or any other type of leader in any organization without being fully aware of what technologies are currently available that empower you and your teams to perform your roles better, faster or more productively.

Don't use the excuse that your IT department is in control (or out of control, as the case may be) of everything and can't give you what you need. That is changing – fast. IT departments are *losing control* because individuals now have more computing power at their disposal than ever before.

Below are some predictions and why I think they are important.

## Your Employees Will Dictate Your Software Choices

Already, many departments within the typical enterprise strike out on their own to buy applications that serve their business needs. This will only continue and spread. In the next few years, we will see entire companies functioning without IT departments because they have purchased software services in the open market that satisfy their needs, in some cases department by department without any enterprise-wide coordination. Will software availability and selection become as easy as dialing a different channel on your FM radio?

That might seem extreme, but there are signs we are headed in that direction. New employees coming into an organization recoil at using the clunky ERP form screens to do their jobs. They want to do their work on their iPhone or Mac Book. Do you think that our young professionals are just spoiled by today's easy-to-use technology? That they aren't smart enough to use more complex forms?

The answer is "yes" to the first question, but "no" to the second. Our next generation isn't dumb or incapable of wading through a ridiculous series of SAP screens, exporting to Excel when needed, and coming up with whatever information they need to do their job. They are just telling us that *this process is stupid.*

They recoil from standard SAP screens because they are used to icon-touch-screen-based computing through an iOS or Android device. Are you hiring them for their ability to navigate SAP, Oracle, JDA or Infor software routines, or are you hiring them for their analytical minds, creativity and imagination, business potential or functional abilities?

Young professionals don't have the patience to wait for an IT department to give them the tools they need. They are immediately full of suggestions for making software work better. They write SAP macros to simplify their work, use dashboard software as a front-end transaction menu, pull applications off the internet, write their own SQL and Microsoft Access queries, and try things they learned from online user forums. How are enterprises possibly going to hold back such drive and creativity?

## Traditional ERP Continues To Look Unnecessary

I don't mean that *the concept of ERP* will become unnecessary, just the traditional ERP on the traditional platform as we have known it for the past 15 years.  In the future I imagine the following trends will bring about this prediction:

- Enterprises will no longer be willing to spend huge sums for year(s)-long projects where they end up owning, hosting and caring for software that requires upgrades every 18 to 24 months, not to mention the expense and management of IT staff to support the data center, integrations, system backups and nonstop user demand for new features and changes to the software.

- Organizations will trend away from capital investments in software and toward paying for software as an operating expense. This is because: 1) a large capital project is much harder to sell than a relatively small long-term increase in operating expense; 2) associating software expense directly to the function that uses it is a better accounting practice; 3) in an ROI calculation, the added operating expense of a SaaS model only has to be offset by yearly savings brought about by the new application, instead of requiring that three or so years of annual savings pay for the entire cost of the new system.

- The software market will continue to offer new products and services that, while not a complete ERP system, could replace large components of an on-premise ERP.

- Companies will begin to piece together "point" solutions – single-purpose applications – in the cloud as SaaS solutions. By networking their applications in the cloud, they will achieve better functionality, far greater flexibility and lower cost than a traditional ERP system.

## Better Options Replace Traditional Data Warehousing

This isn't a novel prediction.  Traditional data warehouse projects are overly long and expensive, and ongoing maintenance is resource-heavy. Why endure these negatives when there are other, less painful options?

The biggest game changer here is that with new capabilities available, data doesn't have to be extracted, transformed and placed in named data tables in order for it to be available for analysis. Instead, it can be viewed wherever it is via a data virtualization tool.

Traditional data warehousing is like putting data in locked drawers. When you update the data you unlock the drawers, put the data in, then close and lock the drawers. Data virtualization is like taking the data out of the drawers and spreading it out on a huge table where you can see every piece of data. In that form, you can quickly pick and choose what pieces you want to put together temporarily.

I like to think of virtualizing the data in an in-memory environment as like suspending the data in midair, everywhere, so that you can create data combinations and leave them dangling out there for future use. *Visibility and quick accessibility are the key advantages.*

The relational database itself – where data is stored in columns and rows – is being challenged, by newer database and search technologies that can sort through much more data, at a faster rate and at a lower cost. This is accomplished by distributing the data sorting tasks to multiple servers, and is referred to by some as NoSQL.

Software giant Oracle owes much of its explosive growth since its founding in 1977 to the relational database. Oracle has long owned a large share of this market, but is today facing competition from firms with less expensive and faster implementation models that is slowing its growth and eating into its core business.

Writing in *The Wall Street Journal* recently, Michael Hickins reported that "Facebook Inc. and Autozone Inc. use NoSQL databases to run all or part of their business processes....other companies have embraced NoSQL databases to store and analyze data as disparate as email, transcripts of call center conversations, and tweets."

**Look for vendors who understand this paradigm shift and who are prepared to deliver the next generation of analytics tools.**

## In-Memory Computing

In-memory computing is another technology leapfrogging the traditional data warehouse. An in-memory architecture uses data that is in the main memory (also known as Random Access Memory, or RAM) of a computer, rather than data on a hard disk. Data retrieval from a disk is the slowest part of any analytics query, because the software has to "find and fetch" the data you want, and queries accessing very large amounts of data just can't be done in a feasible amount of time. You've probably already experienced this. I work with people who run some SAP queries that take an hour or more to run. These people would like to query even larger amounts of data but don't even bother trying because they know SAP might just stop in midstream or take so long that the information isn't worth the effort.

An in-memory setup eliminates find and fetch because the data isn't even stored on a disk; it's available right there in the main memory of the application. As I noted earlier, I like to think of in-memory computing as having everything I need floating in the air in front of me, ready at any time. I can pick several pieces of data to compare, add more pieces as needed, join data together and leave it in-memory, come back later and use the joined data to add to something else.

With current capabilities, a whole database of record for a software application can be stored in-memory. Fast hardware evolution and the declining cost of memory have made this possible.

You might be yawning and thinking "faster queries – so what?"

**This means that you, as a professional in a non-IT function (or even in an IT function) don't have to settle any more for yesterday's analytics solutions**. You, as a more informed professional, can now tell your IT department that you want to explore what is now available, that is less costly and much easier to implement than traditional business intelligence or reporting setups.

It also means that the way you collect, sort, analyze, chart, use and interpret data should change dramatically – from a fixed and limited process to a more natural and iterative process. The in-memory technology makes it possible to gather information a lot like your normal thought process. Your brain is like an in-memory computer. To make a decision, you first start with the information you have in your head. Then you gather what is missing, using the web, asking questions, reading the newspaper. Your brain immediately processes each new piece of information and sometimes in seconds you've made your decision.

**This new paradigm – massive data storage connected to super fast computing power – will change what we ask for.** No longer will we ask for a report on sales by customer, by date, by region, by product. Instead we will want every single piece of data related to any sale of anything to anyone, say, for the past two years–every single invoice, credit, return, price, discount, the person who sold it, the commission paid on it, the color of the product, the shipment date, delivery data, invoice payment amount, date of payment – everything. This will become the expectation in all areas of an enterprise.

## The Visualization Of Data

Technology will make it easy to index every single event in your enterprise and to display in real time a visual interpretation of all of those events interacting with one another. Executives and managers will no longer look at static tables of numbers or even graphs or charts; they will be able to "watch" their business in real time and see a future visualized picture of their business, much like a weather forecast is shown in graphical terms.

This doesn't mean numbers aren't important; numeric values or percentage changes will of course accompany visualizations of data. Visualization is just the next step in presenting events or data in a way that is analytically logical and much more revealing to the viewer.

An example: Frank Luntz is a professional pollster who uses visualization to show the sentiments of viewers as they watch political ads. The technique uses a moving graphical depiction to show *when exactly during an ad* viewers felt positive or negative toward the content of the ad. The combination of the sentiment and the ad reel is what creates the value.

Mr. Luntz could have simply had each viewer fill out a questionnaire about the ad – what did they like and what didn't they like? You would then see numeric totals and percentages related to each question, but you wouldn't see exactly when during the ad viewers had positive or negative feelings. The second-by-second gathering of data draws a much clearer picture.

## A Large Menu Of Choices

Corporate software has gradually become democratized. This means more and more choices are now available to more and more people.

So, after you have defined your basic system requirements, you must next take this very important step: Go shopping. Yes, you should hit the Internet and shop for software. Read the white papers, blogs, industry studies – get information wherever you can. You can hire a consultant to help you if you want, and in many cases this is a smart thing to do. But these days, if you have done your homework, it's not hard to create a short list of software options.

Remember, you are a professional of some kind – finance, management, human resources, operations, sales or marketing – first and foremost, but it's no longer possible to be any kind of professional without being fully savvy about the software tools that support your profession. That makes you an informed shopper.

Example: It is now possible to send your enterprise transaction data, automatically every day, via flat files to a vendor's cloud server, and use the vendor's very robust software (hundreds of preconfigured analytical routines) to analyze the data, with no upfront costs, on a pay-by-month basis, with no annual contract.

FusionOps is one firm that offers this type of business intelligence solution.

The large menu of choices will make it possible for you to put together your own "suite" of systems, even if each of the applications is in a SaaS cloud somewhere. You can hire an infrastructure-as-a-service (IaaS) company to host everything, and one or more SaaS business intelligence services to give you all the analytics you could want. The democratization means companies will now use cloud or SaaS software as an extension of their IT departments. It will be a hybrid world. The days of having to run everything on a particular kind of server using a particular operating system are over.

Nowhere is software choice being more vividly displayed than in apps for smart phones. Between Google and Apple, more than 800,000 apps are now available. Yes, these apps are by and large consumer-oriented, not business applications. But the market is flooded. The world doesn't need 800,000 smartphone apps, which means that software developers will look to other markets; business applications could be the next market for an explosion of mobile applications.

I can see corporate software becoming a collection of appliances, like a kitchen. Each appliance runs on its own, and performs a specific function, yet the output of one is compatible as the input to another. When the next generation of appliances becomes available, you replace the existing one to take advantage of greater capabilities, lower cost or both.

## Muddy Waters In The Software Market

It's getting harder to determine which software vendors have what capabilities. This is because:

- The number of technology startups has increased;
- Big software companies have been acquiring other firms to increase the breadth of their capabilities;

- Established firms are rapidly making changes to their suite of applications – adding capabilities so quickly that it's difficult to land on a static evaluation and comparison vs. other vendors.

- The branding of specific functionality continues to proliferate. Firms don't define their software's features all the same way – they give them a brand name, which only adds terminology that is unnecessary and gets in the way of a clear comparison of features.

Firms are offering products and services that overlap what other firms offer, making it more difficult to weed out who truly offers what you want. (Use the Vendor Checklist in the Appendix as a guide).

It used to be that, if your company needed software in some form – packaged or custom – it was "installed" on a server. Then a "client" for the software – a relatively small piece of software – was installed on desktops so that the software on the server could communicate with the user on the desktop. In between the two was a local-area network (LAN), which is jargon for a wired connection. In this configuration, a user could launch the client software on a PC, and the client would, via communication over the LAN to the server, enable the user to fully use all the features of the software.

Players in this market looked like this:

- The firm that wrote the software;

- The firms or independent consultants that support the software;

- The firm(s) that helped you to install, configure, test and launch the software you bought;

- The company from which you bought your servers;

- The company that supplied your LAN and wide-area network (WA)

All of this has changed. Now there are vendors that can do all of the above, without stepping inside your building, through a web portal.

## The Elusive Breakthrough

No matter how cool and advanced software becomes, the fact remains that organizations still struggle to implement it and make it work so that predicted benefits are realized. Computer languages have evolved and have become much more sophisticated, yet organizations such as the North Carolina Department of Health and Human Services and the U.S. Department of Defense are wrestling with runaway custom software projects. Run-of-the-mill ERP projects still cost millions of dollars and take a year or more to complete.

The elusive breakthrough in how enterprises adopt new software hasn't happened yet because it's so elusive. We haven't figured out how to convert human thought to computer instructions without spending months or years at it, along with horrific amounts of money. That the software industry has grown so much in spite of these hurdles is amazing in itself.

It's clear that big changes are taking place in nearby fields. Just a few years ago iTunes was on top, but today seems almost clunky compared to the ease of using Pandora or Spotify. Who needs a Garmin GPS when you have an iPhone? Does anyone call an airline anymore to buy a plane ticket?

In the same way, I believe SaaS applications will continue to chip away at traditional on-premise software, simply because the market has figured out how to satisfy enterprise software needs with a leaner, less costly and more predictable delivery model.

But SaaS applications don't necessarily solve the larger problem of the cost, time, and complexity of integrating new software applications into your business. This problem is much bigger, and involves, maybe, a whole new way of thinking about how software is developed and delivered. And I don't mean incremental improvements but a real breakthrough – something like being able to capture every single step of every business process in your enterprise, along with all of your master data, in a single electronic file, upload it into a new application, and go.

In the meantime, stay current with innovations in your area of business. It doesn't take a lot of time – subscribe to some of the free industry newsletters and blogs, go to a conference every once in a while, and ask a lot of questions.

The next chapter is an executive summary of the entire book.

---

It was 1995, and I was 35 years old when I started to work on my first big ERP project. By that time I had nine years of business experience including roles as a plant financial controller and as manager of a large distribution center. I had an MBA in finance from a prestigious university. I thought I knew everything.

But on that first project I started out by spending several months basically re-writing what the software should do for my area of responsibility. I drafted pages and pages of documents outlining new logic, algorithms, screens and reports that the software should have. I thought what I had written was pure brilliance. Not only did these proposed modifications turn out to be prohibitively expensive, but the software firm had no interest in making them because they were so far from the base logic and code of the package they had developed.

That was the first rabbit hole I went down. I've stepped into a lot of potholes in the years since 1995, and I don't want you to step in the same ones, which was the purpose for writing this book. If by reading these words you've only just changed your attitude or approach to enterprise software, then I've achieved something.

Hopefully by now you've seen how money pits don't have to happen. Avoiding them is a choice, involving the right attitude toward software, the right steps to follow as outlined in this book, and the confidence and resolve within you to fight off the always-present forces pushing your endeavor toward that financial black hole.

# 13
# ESSENTIALS
# TO WINNING –
# A SUMMARY

This chapter summarizes the main themes of the book, and follows the book's chapter sequence.

## Understand The Minefield

Business, or enterprise, software has a rocky history of success going back to the very first efforts to create software in the 1950s. Software is a collection of computer instructions ostensibly reflecting what the human mind wants the computer to do. The translation of human thought to computer instructions is not easy, and is only precise and predictable in the most simple of calculations. But the human mind is capable of more complex imaginations; unfortunately replicating these thoughts in software form has been failure-prone.

From the beginning, software endeavors have seemed to take a ridiculous amount of time – in some cases years – to complete. This is because software is like no other product on earth. Even the simplest programs may have tens of thousands of lines of code. Programmers are human and therefore make mistakes. It is impossible sometimes for a programmer to even remember what he or she did on a given program for a particular purpose.

Putting new software to work in your enterprise requires the hiring of expensive people for a period of at least several months and in many cases one to two years.

The landscape is littered with large and small failures – epic ones like the Denver Airport baggage system and smaller ones like a collection of ERP enhancements costing $600,000 from which only few benefits were gained. Studies show that well over half of software projects fail – they don't meet stated objectives, take way longer than expected to complete, cost way more than expected, or are not completed at all. This doesn't portend well for the typical enterprise striking out on a large software endeavor, and should make you extremely cautious in your decisions.

Buying enterprise software in the 21$^{st}$ century is an adventure. Software marketing is confusing and fraught with lingo and claims that are hard to verify. Estimated costs are just that – estimates – and are more than likely to be exceeded by the time the project is complete – if it is completed at all.

The enterprise software industry is a collection of players, all of whom have a keen interest in the perpetuation of spending on software and related services. Software and hardware firms, consultants, research and advisory firms, hosting services, and network providers all participate in endless talk, conferences, white papers, demos and industry confabs to keep the topic of investing in new capabilities top of mind among CIOs and senior managers.

You can cut through this fog by:

- Getting quickly past the introductory sales pitches;

- Watching for the inevitable setup of here-is-the-big-problem-faced-by-companies-today-and-here-is-the-software-to-overcome-that-problem;

- Insisting that the software vendor connect the dots for you by showing in detail how the business needs of your enterprise in particular are met by that vendor's solution;

- Having the vendor state in writing how the application meets each one of your specific business requirements;

- Getting an objective comparison of software solutions by a reputable firm.

Experience has shown that it's easy for projects to end up in a money pit. Losing investments are characterized by:

- Unclear ROI or payback based on unrealistic or false assumptions;

- A complex, unproven software solution;

- Technology that its intended users cannot understand;

- A weak project team and inadequate involvement of end users;

- Unclear or continually changing requirements;

- Inadequate vetting of the software, leading to delays and extra costs;

- Big bang cutover without adequate risk management.

Dodging the money pit, while not easy, is entirely possible. From case studies of winning projects we know that they have these characteristics:

- Projects with an unambiguous and demonstrably positive ROI based on realistic assumption;

- Technology readily available in the market and proven to work in other enterprises, with limited customization;

- Full under-the-hood evaluation of the fit of the system with the business;

- Narrow, defined scope of implementation, clear and unchanging business requirements;

- Strong, experienced, focused team with a mandate from top management and a simple, clear objective;

## Nip The Losers In The Bud

One way to leave the money pit behind is to nip losing ideas and losing projects in the bud. You can do this by making sure your enterprise is investing for the right reasons, among them:

- A clear and demonstrable ROI;

- Growth that is outstripping existing systems;

- Regulatory or customer requirements that can no longer be met;

- Too much reliance on one or two people for knowledge and support of your systems.

You can ensure that your enterprise makes the right decision by:

**Positioning yourself as gatekeeper** - *Keep asking questions.* This has the effect of exposing and testing the initial rationale for a project. Don't take sides, just stick to the original business rationale for the investment and keep asking whether the solution being considered matches your enterprise's requirements. Be the voice of objectivity and reason and you will have a huge impact on the success of the effort.

**Evaluating everyone's assumptions** – Below are some of the most common accepted truths that, when not vigorously questioned, can torpedo a project:

- What your company saw in the demo or pilot will work in the real world.

- The software will meet all the business requirements specified before the project started.

- The team won't have to make any customizations other than what was already identified.

- The system's users will quickly learn and accept the new system, and the project will be completed on the promised date.

**Connecting the dots for everyone involved.**   Make people understand that it's a combination of process, tools and people that will produce the win. *If all the dots aren't connected, it's not a win.*

## Step Firmly Into The Future

**Spend a lot of time in the future, determining exactly how you want to use the new technology**, what your business processes will look like and most important, how you will achieve the expected financial benefits.  Don't do this alone, but with everyone who has a stake in the project's outcome.  If they don't have the time to participate in this, cancel the project.  One way to find the money pit is to design the solution and the future processes without including the people you need to make the effort successful.

You can use simple tools to draw the future – whiteboards can be used to show the flow of your business processes and your data (yes, you need to get this detailed).  Here are the key steps to defining the future:

### Preparation

Get your business and subject matter experts and key managers or other stakeholders in a room, focused for a day or more.  Outline the agenda: "The purpose of our meeting is threefold: 1) to define as a group exactly those business problems we wish to address and how they impact our business; 2) Imagine what the future looks like in as much detail as we can, including how specific business processes would change; and 3) quantify the changes we seek in our business measurable – costs, sales, headcount, customer service, and so on."

At this point you are not defining what the software or other technology must do. The job of technology is not to simply exist on its own; it's to enable your efficient and profitable business processes.

This is a highly interactive session best executed with visuals where possible – whiteboards with illustrations, lists, concepts, etc. I once participated in a similar visioning exercise, where groups were asked to draw the future without words. The artists won the day, but the idea was to paint a picture so vivid it instantly captured the future state you desired, and was clearly understood by everyone else in the room.

### State the Problem

1. What is wrong or suboptimal – what is broken, what capabilities are lacking, or what opportunities exist that you can't go after because something is missing? "Customers want our products re-packaged into special formats to sell in their stores as unique offerings…our systems aren't set up to manage orders and invoices for this re-packaged product."

2. What is the impact of the problem? Who does it affect – customer, supplier, employees; how does it affect performance – service, costs, efficiency; and what is the lost opportunity – sales, market share, business growth?

3. In each case be specific. "We need more flexible systems for managing customer orders" is not specific. "We cannot change anything on a customer order without deleting and re-entering it" is specific.

### Imagine

1. When the problems identified in the first step are resolved, what does that look like? What would it look like if, for example, the problem of having to delete and re-enter a

customer order just to change it were fixed? "Service reps can retrieve from the system a customer order, go into an edit process, make any change required, including products ordered, quantities, delivery dates, method of payment, and shipping options, then save the order."

2. Don't be constrained by today's business situation, type of system architecture, or even the preferences of senior management. Don't limit your thinking to just small problems. Imagine a bigger picture, years into the future, when the company is transformed from what it is today...what would that look like?

3. Start with the big picture – "Sales have doubled" – and work downward to specifics from there. "We started selling through a web portal." The big picture is the eventual outcome you want, the specifics – a web portal, for example – will be the things around which new business processes will be formed. The new processes are what you want the new technology to enable.

4. Define the new processes. A 'process' is just the sequential steps that lead to a particular outcome. The easiest way to define a process is to ask questions. What is done first? Then what? Then what after that? You may end up with three new processes, or 20, depending on how big your project is.

**Quantify & Specify**

1. Determine what the future state is worth to your enterprise. What costs do you save, what new customers do you gain, what new products or services can you now sell? Here you need real dollars, not just imagined benefits and rough guesses. Also key performance indicators such as cost per pound, customer order fill rate, and % obsolescence and write-offs, for example – those numbers that you use to track the health of your business.

2. Zoom in on each part or step in your future process and list the requirements of the new software. Example: "enable lookup of a customer's last order," or "display the location of the customer on a map using street address, city and zip code."

3. Log requirements into a spreadsheet file and track the progression of each throughout the project (open, in progress, enabled, and "gap" – the software does not meet that specific requirement).

4. Use the requirements file to evaluate alternative system solutions. It's easy to calculate a "% fit" using your requirements list – if you have 200 requirements (not unusual) and the software appears to meet (really meet) 165 of those, you have an 83% fit.

## Have A Strategy!

Before leaping into a software investment, determine *what your company's general approach is* to making these kinds of decisions. Examples of different strategies:

- The complete outsourcing strategy: "We outsource all of our IT and software needs. We don't own any applications; we just use them as a service."

- The pure ROI-based strategy: "We base our decisions on software investments according to strict financial hurdles. Every investment has to pay back within 36 months, and we include all external and internal costs in that ROI model."

- The anticipated growth strategy: "We use ROI as one measure of whether the investment should be made, but we also heavily weigh anticipated future needs."

- The complete in-house strategy: "Our business model is so unique, and our processes and competitive advantage so specialized, that we mostly have to develop completely custom software."

**Make the strategy fit the model of your business or enterprise.**

- Align your software selections with who your company is and where it's going.

- Invest in applications that help you do excellently those things that *are critical to your business.*

- The packaged application options available to you depend on what part of your enterprise will use the application, and what you want the software to do.

- Completely custom software can be part of your strategy if what you need is unique to the core success of your business.

Tread carefully when considering custom software. Custom software projects are more risky, more difficult to estimate and more unpredictable in terms of how long they will take to complete.

**Some small parts of a project are good candidates for custom development**. Integrating two or more applications, for example, via custom data transfer mechanisms, EDI or some other way, is an activity well-suited for custom development.

**Understand what the packaged software market is offering – it changes frequently**. Whatever your role is in an enterprise, it pays to know what new applications are available in the market to help you do your job. If you don't, you may be buying yesterday's technology.

**You will find that many software vendors are trying to grow beyond their core expertise by claiming their solutions can also handle other functions within an enterprise.** Be careful about this:

- While a package can indeed perform other functions, it is likely to do so with limited features;

- There are usually other vendors that specialize in those other functions;

- If you just want the core functionality that the application was originally designed for, you'll need to determine how to use just that portion of the solution while integrating it with the rest of your enterprise's software.

**Stick with a software vendor's competencies**. Unless you see hard returns in expanding beyond the package's main mission, stay within your scope.

**The biggest risk in overspending with packaged applications is during the implementation phase**. A single implementation of, say, a system to manage just order processing, can cost $2 million to $3 million or more, because you don't know what modifications are needed and you don't know what delays you might encounter.

**Plan the life-cycle costs of a packaged application**. These costs include demands from users for more features (custom development work), annual support fees from the vendor and anticipated vendor upgrades – the cost of moving from one vendor-supported version to the next.

**Understand that software is an annuity business.** In simplistic terms, a software company survives in the long run because it is able to collect annual maintenance fees – typically 20% to 25% of the original software license cost – from its customers while providing custom development services and a stream of version upgrades. This means you are tied to that vendor for at least several years, and will effectively be paying the full cost of a new license every four to five years.

### Be smart about integration

Nearly all new applications you add to your enterprise will need some level of *connectivity to your existing systems*. Don't ignore this aspect as being a technical detail not worthy of your attention.

Be familiar with these two terms:

- Application programming interface (API) - An API is a protocol used by software components to communicate with one another. Ask software vendors whether they have developed APIs for interfacing with other programs, and which programs in particular.

- Electronic Data Interchange (EDI) - is a standard for electronic messaging of commerce between two entities and frequently between two different systems. Don't underestimate the time and cost of integration via EDI, including ongoing support costs.

**The quality of your master data will directly affect ease of integration across your enterprise.** For applications to communicate with one another, they have to use the same definitions of data. If your master data is not standard across your enterprise, the extra code necessary to translate incongruous data will add to the complexity and cost of integration.

To further control costs and complexity of integration:

- Use a hub structure if multiple applications need the same data.
- Document everything. Don't end up with a complicated setup that only one person is familiar with.
- Let your ERP system define your master data standards, and keep those standards consistent in all other systems.
- Keep the number of integrations to a minimum and the integrations themselves as simple as possible!

**Have a structure for evaluating your choices**

You and your team should ask the following questions when evaluating a software solution:

- Does it solve my problem?
- Does it pay back?
- Do I understand all of the solution's costs?
- Is the solution in line with my strategy?
- Do I understand all of my alternatives, besides this particular vendor?

- Does my team have the time and skills to implement this solution?

- Do my users have the aptitude to learn it and become proficient?

- Does my team fully understand how this solution will integrate with the company's other systems?

- How risky is this particular software alternative compared to others?

- Is the vendor's reputation verifiable?

- Can I find programming help in the open market?

## Understand Basic Software Choices

See Chapter 9 for an extended description of these parts of the software market:

- Big ERP and other ERP – the leading players in end-to-end multi-function software for all parts of your enterprise.

- Manufacturing – applications that manage production and assembly processes.

- Supply chain – software for demand planning, inventory management, transportation, warehousing and order processing.

- Finance – accounting software and the accompanying analytical functions.

- CRM and sales management – applications for analyzing and managing customer performance and programs, and analyzing impact of selling strategies.

- Analytics – software for sifting, sorting, visualizing and extracting insights from data.

- Big data and data virtualization – the latest developments in software that can crunch staggering amounts of data from disparate sources.

- Enterprise information management – software for indexing, searching and compiling information (not necessarily data) from all of the places in your enterprise where documents might reside.

## Consider Saas, Cloud And Mobile

### SaaS and Cloud

**Companies with limited IT resources should always consider a cloud or SaaS solution.** What is new is an expanded range of products and services offered via a cloud platform – today some companies can run virtually an entire business in a cloud environment.

**Inexpensive cloud applications are available if you can run your business with plain vanilla (out of the box, no customization) applications.** Enterprises can access what is called a multi-tenant version of the software – several companies use the same instance, but whose transactions and data are separated from one another by functioning in a different location or 'node' of the software.

**With so many options now available in SaaS form, it only makes sense to explore these alternatives.** The benefits are: no server maintenance, no operating system license or maintenance, easy access via the internet, telephone support, custom development options, and near-seamless transitions to newer versions. The downside is a lack of control over things like planned outages, and a reliance on a third party for mission-critical applications.

### Mobile

Employees frustrated with corporate-issued Windows-based PCs are increasingly asking to be connected to their companies via iPhones, iPads and MacBooks. But mobile devices weren't necessarily designed to connect to big enterprise systems, so if the promise of mobile is to be realized, new applications will have to emerge to enable this connectivity. "Re-platforming" from desktop/PC to mobile will likely mean a wave of new applications hitting the market.

Mobile computing is not just about convenience, and you should care about this trend because you're likely to come across some high-ROI investment possibilities.

If your enterprise is considering significant investments in mobile computing, here are some things to keep in mind:

- To capitalize on mobile you need software on the device and a way for the device to reach your enterprise systems.

- Put in place some type of governance around requests for mobile access.

- A decision to invest in mobile applications should involve the same rigorous process as any other software investment: Step firmly into the future; pick the right technology, and manage ruthlessly. To keep risk low, pick a small demonstration project.

- Mobile isn't just for convenience. The people who interact with your customers or who work in the field need data, analytics and the ability to transact at their fingertips.

## Manage Ruthlessly

**Create a ruthless team**. The strength of your team is by far the most important factor in creating a winning investment. If you cannot form a strong team to carry out an important and expensive software investment, you should not make the investment.

**Your project manager is someone who is expert on the business processes you are trying to change,** and preferably also familiar with the systems currently used to enable those processes.

**Your project manager needs two experienced and competent partners**; 1) the most senior and experienced IT person in your enterprise whose functional (business) expertise matches the nature of the project; and 2) an experienced implementation professional, preferably from a leading technology services firm. **Give**

your project manager the benefit of a partner who has done this before, preferably many times before!

Your job is to ensure your (strong) PM gets the needed resources.

Give the team the mandate in the strongest possible terms. Keep the mandate simple: "Your mission is to move these five business processes to this application, and in the process fully realize these three main business benefits."

### Embrace structure

Corporate-casual norms are anathema to a serious and successful software investment project, which demands rigor and, above all, structure. Structure in terms of:

- A work plan with tasks, due dates and responsible owners that is reviewed daily;

- System requirements, testing scenarios and future processes that are documented in detail and agreed to by all stakeholders;

- An issue resolution process to ensure impediments to progress are eliminated;

- Commitments by team members to own and deliver key parts of the project on time;

- An overall sense of urgency to complete tasks and kill barriers to progress;

You need a structured methodology, but not necessarily the one your software vendor or consulting partner uses. A methodology is simply a way of doing things.

Don't burden the team with excessive administration. Every project needs a traffic manager with good tracking tools who is in control of the project plan and is monitoring the status of each event in the plan, but don't go crazy documenting everything in sight.

Make people pay attention!

**It's not how many people you can get on your team, but how much of their focused attention you can get when you need it.** Attention deficit will derail your project just as sure as any other threat.

### Kill the project killers

*Project killers stem mostly from the decisions made by management or by the project team.* Among them:

- Selection of incorrect technology or software that needs way more customization than you thought;
- Unrealistic deadlines for implementation;
- An overly complex solution design;
- An inexperienced project team or lack of knowledgeable people for the project;
- Slipping deadlines leading to a late project;
- Lack of user or business expert involvement.

### Learn from your own experience

**Resist the temptation to call your project a success** and instead take a hard look at what went wrong and why. Even very successful projects can stand some improvement. And since the overwhelming majority of projects are flawed in some way, you may as well learn from the failures because you and your enterprise will most certainly invest in software in the future. What will you do differently on the next project?

**Did you really get the ROI you planned?** Not what you would present to your boss' boss, but what did you *really* get, compared to your original business case? A clear-eyed acknowledgment of both the gains you got and fantasies that didn't come true will help you (next time) to avoid making one of the classic IT investing mistakes – fooling yourself about the ROI.

**What do the users think about the project?** At the risk of unleashing a firestorm, you need to get feedback from the application's users. Separate complaints from useful feedback and file them away for future use.

**How did your software vendor(s) and implementation partners perform?** Would you use the same people and firms again? Take out the contract and see if everyone did what they said they would do.

# Appendix

**Ten Easy Ways To Land In The Money Pit**

Sometimes in a how-to-succeed book it's enlightening to see *what not to do.*

It's quite possible that there are as many ways to derail a project as there are projects. Each project that fails has its own unique set of characteristics, timing, and decisions that led to failure. What follows are 10 ways you can torpedo a project, all of which I have experienced at one time or another.

1) **Make the decision to invest in a new system for the wrong reasons**. A classic mistake, for example, is thinking that new software will fix a broken or inefficient process. Companies launch into a project trying to fit their broken processes into the new application, when what they could do instead is fix the broken process and derive benefits from that.

But fixing processes is boring – few people want to take the time – instead a silver bullet is imagined, in the form of a new system.

2) **Fool yourself about the ROI you'll achieve with the new system.** An ROI for a ham sandwich! Yes, it's possible, which is another way of saying a project ROI always has to be viewed with a degree of skepticism. It's easy to overestimate benefits and to underestimate costs and time. This is especially true in a corporate environment, where the CFO is usually the gatekeeper for

projects that are approved, and everyone knows they need good numbers to justify what they want to do.

3) **Choose the wrong technology**. Decide to launch a big ERP project when all you really need is a sales planning and trade promotion system, or an updated financials and payroll application. Develop the software yourself even though there are reasonably good packaged applications available on the market, or select a vendor with a cutting-edge application that no one has yet implemented; or go with a traditional on-premises client-server solution because you don't know about other options available (which would be cheaper and easier to implement).

4) **Assign the project to people who are already swamped and who live in different time zones**. Run most of the project via conference calls to save money on travel. (Remember, one of the keys to success is full engagement in and focus on how the software will enable the business to run. The team's on-site collaboration is essential, so if you **don't** want the project to fail, eliminate as many roadblocks as possible to getting the team together in one place).

5) **Put your weakest people on the project team**. Select a project manager who is leading a project for the first time. Don't free up the critical subject matter experts that will be needed; they'll find the time somehow. Just empower the team and everything will run fine.

6) **Make sure the application is fully modified** to match every nuance of your business so people are comfortable with it. Don't worry if the solution becomes too complex – it's software so everything can be programmed.

7) **Don't place a fixed deadline or budget for project completion**; plans need to be flexible to account for unforeseen difficulties, or opportunities for expansion of the solution to satisfy more of your company's needs. A fixed deadline might be too risky if the software isn't fully ready to be launched.

8) **Make sure everything is launched all at the same time**; the solution may not work properly otherwise. Breaking up the project

into parts or phases tends to drag the project out, and temporary system interfaces or manual processes may have to be used until the full solution has been installed.

**9) Don't assign anyone to the team who would be one of the future users of the system.** The application's users need to be trained at some point, but their involvement in the project early on might disrupt focus on the core mission of implementing the solution. Since the project team has to completely focus on making the software work, they can't worry about the user population and the organization as a whole being ready. There is always a learning curve, and people will gain proficiency over time.

**10) Don't try to manage the project with a structured methodology.** Doing so will unduly restrict the free flow of ideas and creativity that are essential. Most project methodologies are a waste of time and only serve to increase the workload on the project team.

## Lingo Translator

Confused by tech-speak? Here is a partial list of terms, what they mean, and how they might be used in a conversation.

**Table 4**

**Lingo Translator**

| Lingo | Meaning | Example |
|-------|---------|---------|
| **Application** | A software program designed to perform a specific function | "XYZ is an application for determining optimal inventory levels." |
| **API** | Applitcation Programming Interface. Software code or code specifications that have already been written whose purpose is to transfer data from one application to another. | "Company B says they already have APIs for integrating to the main packaged ERP systems." |
| **Architecture** | The structure, as envisioned in a diagram, of a company's systems, typically showing applications, databases, interfaces, networks, file transfers, and user interfaces. Shows the flow of data and transactions. | "The application they demonstrated looks good but we need to understand the overall architecture, especially how it would fit in to our existing framework" |
| **Bandwidth** | Two uses: 1) the data transfer rate (or 'speed') of a network, usually expressed in megabits per second (mps); and 2) ability, time, or capacity of a person or persons to carry out work. | "If the application will be hosted outside our network, we will need to increase the bandwidth of our external internet connections." "The team just doesn't have the bandwidth right now to expand the project scope any further. |

| Lingo | Meaning | Example |
|---|---|---|
| **Base Code** | Refers to the collection of code that is common in all versions of the software. | "Can we migrate to the next version, with our custom modifications, using the same base code?" |
| **BI** | Business Intelligence, an all-purpose term for queries, reports, analysis of an enterprise's data, for any purpose - sales trends, cost trends, key performance indicators - any information that helps management understand business performance. | "The company has always relied on internal programmers for BI. They would write special queries or extract data and put it in a form so that the user could browse through it with Business Object or Excel." |
| **BI Stack or just Stack** | Viewed diagrammatically as systems 'stacked' on top of one another, usually with transaction databases on the bottom, a data warehouse in the middle, and business intelligence (BI) or other query tools on the top. | "We are trying to move our BI stack to the cloud - databases, data marts, and query software. We think there are savings in moving these operations out of the data center." |
| **Box** | Used interchangeably with the term 'instance' and 'version,' even though 'box' is typically used to refer to a physical server. | "We think they are finished with the modifications but the programmers want to test it some more on the development box." |
| **Code** | Computer code. A software program's core set of instructions and logic. | "That kind of change will probably require a rewrite of some code." |

| Lingo | Meaning | Example |
|---|---|---|
| **Configuration or Config** | The settings, parameters, flags, selections and other setup a software application needs in order to work. | "We have a test version of the software set up, and it's configured according to how our enterprise would use it." |
| **CRM** | Customer Relationship Management. Software for keeping track of all customer information relevant to the objective of increasing sales and profitability through each customer. Also tracks pricing, promotions, profitability, average order size, and numerous other attributes important to knowing a customer's business. | "Our pricing and promotion deals have become complex and hard to keep track of. We need a true CRM system to make sure we're spending promotional dollars effectively." |
| **Cubes** | Sets of data that are pre-staged or pre-organized so that a browser or other query tool can extract and manipulate the data according to user needs. | "We have set up an invoice cube so that the users can use Business Objects to create their own queries of customer invoice data." |
| **Data Center** | The room or building that houses a firm's computers that run software for the business. | "Last year we moved our data center from in-house to an outside company that will host all of our applications and manage server hardware and software." |

| Lingo | Meaning | Example |
|-------|---------|---------|
| **Data Mart** | A repository of data from specific systems. | "The Finance team wanted a lot of invoice data so we set up a data mart that collects all of the data from each invoice on a daily basis." |
| **Development Instance** | The version of the software that programmers use to modify computer code and test it before transferring the completed code to a test or production instance. | "We think they are finished with the modifications but the programmers want to test it some more on the development instance." |
| **EDW** | Enterprise Data Warehouse. A repository of data collected from an enterprise's transactional and other applications, which is used to create reports and other analyses for management purposes. | "Most of our day to day reporting comes from a data warehouse we set up to collect daily transactional data such as sales, production, and inventory." |
| **ERP** | Enterprise Resource Planning system. Interconnected applications that support and enable the interconnected processes of an organization, according to an enterprise's business model. | "We use an SAP ERP system for most processes, but have specialized applications for customer relationship management and for transportation management." |

| Lingo | Meaning | Example |
|---|---|---|
| **Footprint** | Usually means application footprint, or scope of business processes that the application will enable. | Do you have an application footprint diagram so I can see which processes are supported by which applications? |
| **Functionality** | What the software is capable of doing. An accounting and finance application usually has functionality to compute cost of goods sold and post to a general ledger account. | "We looked at Company G's system and the functionality we needed just wasn't there." |
| **Gap** | The difference between what you want the software to do and what the software is capable of doing. Gaps can be 'closed' by using a manual work around (living with it), or by developing custom programming to modify the software. | "We looked at Company G's system and there were just too many gaps." |
| **Governance** | A management mechanism usually made up of rules that are used to define data structures, system changes, or how IT manages requests. | "We have too many data errors and inconsistencies that cause problems in our transaction systems; we need some kind of data governance." |

| Lingo | Meaning | Example |
|-------|---------|---------|
| **Hadoop** | A free, Java-based method of computing that distributes computing tasks to multiple computers called clusters using simple programming rather than using one large computer and database. The computing model is structured so that it can scale up or down depending on the load. | "It seems XYZ vendor is using Hadoop for most of the application's functionality, and adding a few extra components to make it appeal to a particular segment of the market." |
| **Host/Hosting** | The computer server(s) and/or the entity/entities that manage computer servers that are used to run software applications. | "Our email software is hosted by IBM, and the rest of our applications are hosted by a 3rd party data center." |
| **IaaS** | Infrastructure as a Service. Servers and other hardware that are made available for computing based on usage or time. | "Company Z is changing its IT strategy by outsourcing its data server to IaaS providers." |
| **Infrastructure** | Refers to the hardware and networking setup that supports a company's systems. | "We have a lot invested in our infrastructure - all of our servers and network management is in-house." |

| Lingo | Meaning | Example |
|---|---|---|
| **In-Memory** | Refers to a computer's main memory (RAM), not its disk, or hard drive memory. Through lower costs of memory, the large amounts of data usually thought of as occupying disk storage can now reside "in-memory." Can be used to analyze big data. | "The vendor's pitch is that the in-memory architecture means we can process more data faster." |
| **Instance** | Defined as the particular version of the software that is installed on a specific server. | "We share the same instance of the application as our sister companies, but we can't see their transactions and data and they can't see ours." |
| **Legacy** | Applications, data, or platforms that your enterprise is already using, before it implements new applications or platforms. | "We are keeping our legacy payroll and HR systems and building integration to the new ERP System." |
| **MapReduce** | A programming framework, made popular by Google because it is used in Google's search engine, that breaks a software application down into parts and lets each part run on any 'node' in a cluster of servers. Hadoop was inspired by MapReduce. Both are open-source programs, allowing anyone to use them to create applications. | "MapReduce is an easy to use framework; It lets programmers use their language of choice, and its benefits are mostly in processing large data sets, where a search of terabytes or petabytes of data is needed, as in the case of Google indexing all the web pages on the internet." |

| Lingo | Meaning | Example |
|---|---|---|
| **NDA** | Non-disclosure Agreement that is signed by vendors, consultants, or visitors to a company that binds the signing parties to confidentiality regarding any information discussed or exchanged. | "We can start sharing data and working on a system design with the vendor as soon as the NDAs are signed by both parties." |
| **Network** | Series of connections between computers. A LAN is a Local Area Network typically found inside the four walls of a building. A WAN is a Wide Area Network typically used by an enterprise to connect its geographically separate locations. | "Our network includes a LAN for each sales office, and a WAN that connects the whole company together." |
| **Platform** | Type. Used very broadly. Can mean a type of operating system, type of integration, type of software vendor, type of hosting, etc. | "They are on an Oracle (or Windows, or SAP, or EDI, or legacy) platform for most of their systems." |
| **Point Solution** | An application that primarily performs one function, or addresses one particular key area of a company's business. | "We just need a point solution since there are few users and no need for integration with other parts of the company" |

| Lingo | Meaning | Example |
|---|---|---|
| **Production Instance** | Defined as the particular version on a particular server that will be the 'real' version you will use to run your business. Your enterprise is said to have the software 'in production' when all of your set up and testing is complete and you are using the system to operate your company. | "To set up the production version we will have to upload all of our master data to it and run checks to ensure accuracy." |
| **Release or Version** | Variation of the base code that enables the software to provide different features and functionality. | "What version of SAP are you running? We are on ECC 6.0." |
| **RFP** | Request for Proposal. Something a company might send to prospective software vendors, describing the type of application(s) needed and the specific requirements of that application. | "We'll look at the responses to our RFP to see what options there are among the vendors we selected." |
| **SaaS** | Software as a Service. Software that is made available for use based on usage, time, number of transactions, number of users, or a combination of variables. | "We're considering a SaaS solution to save time and avoid requesting the capital dollars needed to buy and install the software." |
| **Solution** | The desired or selected application(s) configured the way an enterprise needs to meet its needs. | "We're trying to make ABC work as our demand-planning solution." |

| Lingo | Meaning | Example |
|-------|---------|---------|
| **Space** | Segment of the market a software product is intended to compete in. | "There are new apps in the CRM space you should consider." |
| **Spaghetti** | Overly complex connections between systems. | "Our system diagram looks like spaghetti with all the complicated linkages we have built up over the years." |
| **SQL** | Structured Query Language. A programming language that is used to update, delete, and request data from a relational database. Often referred to as a 'sequel query' or 'sequel program.' | "We can set up a standard SQL query that could pull the data we need every day and put it into a report format." |
| **SQL Server** | A relational database management system that uses the SQL language. There are many versions of SQL Server, Microsoft and Sybase both publish and distribute their own proprietary versions. | "We have both of our main data warehouses running in SQL Server from Microsoft." |
| **Systems** | Multiple software programs designed to perform functions separately or together. | "The company wanted to re-evaluate its accounting and payroll systems." |

| Lingo | Meaning | Example |
|-------|---------|---------|
| **Tables** | Normally refers to the tables in a database. A table is used to store data in columns and rows, with each table storing one type of data , such as 'customer zip code.' Each table is given a name to display it and its relationship to other tables on a system's architecture diagram. The customer zip code table might be named 'custzip.' | "Do we know which tables are accessed in the system to display data in these fields?" |
| **Test Instance** | Refers to the version of the software you will use for testing. | "For the test instance we won't need all of our master data, just a subset of customers, products, plants, and raw materials." |
| **UAT** | User Acceptance Testing. Refers to a phase in a software project when the software's future users transact their normal daily business on a test system to confirm the system functions as expected. | "We are scheduling UAT for the fall when most users will be available." |
| **WMS** | Warehouse Management System. Software for managing inventory and workflow inside a warehouse or distribution center. | "Our WMS handles first-in first-out inventory rotation and gives us alerts on product approaching expiration date." |

## What's An ROI?

ROI, or Return on Investment, is essential to making the right decision about enterprise software. In simple terms, ROI is the percentage return you are realizing on the money you spent on the software.

ROI = ($ benefits per year minus additional yearly costs from the application) divided by the costs of implementing the new system.

If you spend $2 million up front for an application and another $400,000 per year to maintain it, and the benefits produced amount to $700,000 per year, the ROI would be ($700,000 minus $400,000) divided by $2 million, or 15%.

What ROI should you get from a successful system investment? That depends on your company's strategy and overall approach to these types of projects. But I would say 30% to 40% should be a minimum. That's because an annual rate of return in this range would "pay back" the original investment in about three years, which is about the timeframe within which better software options may exist in the market.

Payback is another way to quickly assess the relative attractiveness of a system implementation. In simple terms, payback is the number of months or years it takes for the financial benefits to "pay back" the original investment.

In the example above, the upfront cost was $2 million and the net annual savings were $300,000. So how many years would it take to pay back the $2 million? The answer is $2 million divided by $300,000, or almost seven years. Using the three year payback guideline, this particular example would be a loser.

## How Do I Get My ROI?

One of the hardest things to do when investing in software is to estimate a true return on investment. Not a fake one – one that you gin up out of thin air that sounds good and will resonate with management – but a real one.

You can fake anything – as I said earlier in the book you could justify an ROI for a ham sandwich. This is probably the prevailing approach in many enterprises, unfortunately. Enthusiastic people get together and the claims of benefits feed upon one another and before you know it a business justification is trotted in front of senior management, and it is well packaged and designed to hit hot buttons. If you want to drum up a business case like that, stop reading. If you want a real return for your project then this section of the book is for you.

This section will focus only on ERP systems, and is organized by functional area.

### ERP Software

An enterprise resource planning, or ERP, system is a collection of software applications that are all connected to each other so as to mimic and optimize the proper flow of data in harmony with an enterprise's business processes. The applications all share the same master data (customers, suppliers) and are usually already integrated with one another. An SAP ERP system, for example, has numerous modules, which are the function-specific applications inside the ERP, such as IM (inventory management), FI (financials – general ledger, accounts payable, etc.) and SD (sales).

In the beginning, when companies are small, they can manage with Excel spreadsheets. At some point, transactions become so numerous and things like pricing, customer information, formulas, vendor contracts and financial statements become so complex that spreadsheets, even with macros and pivot tables, become too unwieldy, inefficient and error-prone to use any longer.

Where to look for a return on investment for ERP software? Because an ERP system covers so many business functions, there are many places to look. Following are the basic areas in which to look for ROI in an ERP system.

ERP systems vary widely in terms of their functional "footprint." Some ERP systems offer soup-to-nuts functionality, with modules in every possible area to satisfy all your business needs. Others are strong in the basics like accounting, manufacturing, and sales. We'll stick to the basics here.

### Master Data

Master data in an ERP system is the single source of, and the official definition of things like "customer," "product," "employee," "vendor," and "plant." There is one definition of each for the whole company. This means that when a financial report is run on sales by customer, and a sales report on the same customers, the entity known as "customer" is the same in both reports.

**Benefit**: consistency of data across the enterprise and one data entry point for use by all modules within the ERP system. Dollar-wise, the benefit could be in lower administrative labor costs.

### Purchasing

An ERP system can enable you to match invoices with receipts and purchase orders so that what you pay for is what you ordered and what you received. The purchase orders, receipts, and invoices are all in the same system, so the system can compare them and immediately determine if the invoice is valid and should be paid or if it's not. **Benefit**: elimination of overpayments or duplicate payments to vendors, reduction of paperwork and manual comparisons reducing administrative overhead.

An ERP system can also manage your contracts with vendors, including pricing and terms of payment. This means that data from invoices can be instantly compared to contractual terms to make sure the invoice is correct. The benefit is the same as above – elimination of overpayment. If your enterprise is large and is processing a large volume of vendor invoices you are bound to have at least a small percentage savings – say 5% of the amount you spend – and 5% of a big number may be enough to pay back at least part of the ERP investment.

A price-shopping or auctioning application or an online buying service can be an extension of the ERP system so that you can search for the best price for your materials, goods, or services, select the vendor, and place the order. Usually these apps and web-available services are specialized according to what you are buying, such as transportation and delivery services, office supplies, basic materials such as standard corrugated packaging, shrink wrap, paper stock, chemicals and industrial supplies, and more recently energy sources such as electricity and natural gas. **Benefit**: getting the best price and terms and automatically creating a purchase order which is integrated to your financial system for proper payment. Again, the dollar benefit can be a percentage of your total spend, especially if you think you haven't opened up your purchasing to alternative vendors for awhile.

When purchasing is part of your ERP the proper postings to financial accounts are automatically done. When you issue a purchase order an entry is made in the ERP system that authorizes receipt of whatever you are buying. When you receive what you are buying a payable is created which goes on the balance sheet as a liability. All the accounting is taken care of.

### Financials

"Financials" is a term used by software vendors and others that normally includes accounts payable, accounts receivable, balance sheet and P&L. There can be extensions of this definition to include such areas as payroll, treasury (bank account inflows and outflows), and tax management.

In an ERP system, all of the required financial postings are made as other transactions take place. A shipment to a customer generates an invoice and posts the accounts receivable for that customer. Production of finished goods creates inventory with its corresponding value on the balance sheet. **Benefit**: reduction in administrative labor needed to manually post transactions from one system to another.

Because the Finance module in an ERP system records all the operating transactions, that data resides in the main ERP database, which means it can be extracted for reporting and analysis purposes. If the Finance module is "robust" enough, it will already have built-in queries or user-defined reports to analyze the basic transactions such as sales by customer, manufacturing costs by product, and other "intelligence" needed to manage the enterprise. **Benefit**: more information available that is critical to evaluating the performance of the business. Calculating actual dollar benefits here can be difficult, but consider what you could save if you knew things like how much overtime pay you incur and in what areas of the business, how much profit or loss you are trending year to date, and which products generate the least profit margin.

ERP systems usually define authorization levels for different types of users, allowing control of sensitive transactions. The Sarbanes-Oxley law and other regulations require separation of duties to ensure financial controls are followed. **Benefit:** centralized control of transactions users have access to and a system infrastructure that satisfies auditor requirements.

### Manufacturing

The manufacturing modules in an ERP system typically manage the conversion of raw materials into finished goods, recording all the relevant costs thereof, including labor, energy, processing and packaging steps, and posting all of these to the P&L and balance sheet where appropriate. **Benefit:** a way to standardize production steps and their associated costs, and measure actual costs to standard costs for purposes of improving efficiency. This is basic cost accounting and cost management, and many companies still do not have decent systems to enable this. The actual dollar benefits come from knowing which products or which production processes generate the most waste, for example, or incur the highest labor cost.

The manufacturing module receives data from "upstream" in the ERP system, such as how much product must be made to satisfy sales demand. The integration of this data, if done correctly, can streamline the daily production scheduling process by avoiding manual intervention to optimize machine time and labor. **Benefit**: reduction in administrative labor and a tight link between what is truly needed – based on a demand forecast – and what is produced. The latter prevents making finished goods that aren't needed, and in industries where products have a short shelf life or can quickly become obsolete, carefully managed inventory levels an prevent costly write-offs. The dollar benefit here can be calculated as a percentage reduction in product obsolescence.

The manufacturing module also sends data downstream to other modules in an ERP system for purposes of transportation, warehousing and distribution. Warehouses can see in an integrated ERP system how much was produced by the factory, how much is in transit to them, and how much more production is scheduled and when. **Benefit**: the distribution end of the company can efficiently plan transportation and warehousing knowing in advance what is coming to them in the production pipeline. Securing transportation well in advance of need tends to reduce the cost of that transportation.

### Sales

The sales module in an ERP system will typically define, record, and track sales by customer, product, region, territory, and brand; it may also provide some CRM functions – these are features that as a whole provide a complete picture of customer performance, attributes, profit analysis, and sales trends. Firms that have a lot of customer-specific programs such as marketing events or merchandising incentives would benefit from these CRM features.

The sales module can provide a window on who you are selling to the most, who is the most profitable, who is growing the fastest, and who is lagging. **Benefit**: Business intelligence about product performance in different customer situations, which customers might benefit from marketing or sales programs and an evaluation of the impact to sales of new products, advertising, or price incentives.

The sales module will use all of the customer master data that is stored in the main database of the ERP system. This includes data such as customer number, bill-to and ship-to name, address, and contact information, price lists, region assignment, products authorized, customer special instructions, order lead times, discount terms, invoice terms, credit limits, and other important information. **Benefit**: one definition of the customer, applied to and used by all parts of the company. No need for translation from one system to another or from one report to another.

Some sales modules will have features that allow you to manage complicated and costly terms of trade, such as sales incentive formulas or price discounts that depend on the customer reaching a specific sales volume target. Some companies have thousands of these incentives in place throughout the year, and it's impossible to manage them without a good system. **Benefit**: more control over a very large expense, and the ability to see how incentives are affecting sales across different customers and products. If selling expenses are a large portion of your costs, you can use these parts of the system to budget promotional efforts by product and sales region. Overall, tighter management of these costs should lead to at least a small percentage savings.

### Supply Chain

The supply chain – a term used to describe all the activities needed to bring goods to market, is a natural beneficiary of a good ERP system because the supply chain extends from one end of an enterprise to another. Often the individual parts of a supply chain run on their own systems – a system for raw materials, another for planning finished goods production and deployment, and so on. It's possible to have three or four separate systems running different parts of the supply chain. With an ERP system, all parts of the supply chain are connected to one another.

**Benefit**: all of the costs and activities associated with each part of the supply chain are in one place, and they are mutually dependent on one another as either inputs or outputs. It is at least theoretically the logical software equivalent of the connected enterprise. And if the enterprise is truly connected and coordinated, then there is never any wasted inventory, lost sales, out of stocks, overproduction, late deliveries, etc. Reality is of course different, but for now let's accept that premise.

A good ERP system will coordinate the supply chain like this:

- It will have a sales forecasting module that allows users to source historical sales data and model it to derive projected sales;

- The projected demand, by day, week, and month, will automatically determine quantities of raw materials needed, where, and when;

- The required materials will automatically be converted to purchase orders for those raw materials;

- The projected demand will determine a production schedule by plant and a schedule of where the finished goods are to be shipped;

- Sales orders are received, checked, confirmed, and sent to the warehouse or distribution center for shipment;

- Sales orders will "consume" a sales forecast so that managers can track shipments against projected sales;

- Sales orders will be routed for delivery via the most efficient method of transportation and transportation providers will be confirmed and scheduled;

- Inventory will be shipped to distribution centers according to the geographic demand of product;

- Distribution centers will receive sales orders for shipment, and shipments to customers will recorded, posted to financials, and used to generate an invoice;

- Warehousing and transportation costs will be posted back to the financial system and the corresponding accounts payable will be created.

### Human Resources

One of the basic things you should be able to do with a good ERP system is manage the costs of human resources. There are many different HR applications available that will do all kinds of things – recruiting, performance reviews, time and attendance, labor management, and benefits management and delivery. But the first and most basic thing to manage is the costs of fixed salaried and hourly labor because these are required for a full financial picture of the enterprise.

**Benefit**: An integrated ERP system can record costs of payroll and apply these costs to different lines on the P&L such as administrative overhead, manufacturing labor, and sales force costs. The first step in managing costs is measuring them, so just the simple act of recording and classifying labor costs usually leads to better cost control. Other systems that manage benefits or time and attendance can be integrated to your ERP system so that dollars in one system equal the dollars in your ERP system, which will be your "system of record" for official reporting and tax purposes.

### Real ERP benefits

Below are some examples of actual benefits I have seen achieved through implementation of an ERP system. This is a miscellaneous list, derived from my experiences on projects.

- Spoilage allowance management yielded about $6 million in annual savings. A company I worked with implemented an ERP system with financials, sales, demand planning and manufacturing. The sales module allowed an overhaul of customer terms, with more specificity by customer. The system enabled an allowance program for product spoilage that replaced the company's prior practice of processing spoilage claims separately as customers submitted them. Customers were happier because the allowance could be profit for them if they managed their product spoilage more carefully.

- Increased forecast accuracy with an integrated demand planning system. In this example, we implemented a fore-casting, product deployment, and manufacturing module as part of a larger ERP system. The demand planning model replaced a clunky and inaccurate system with features allowing more precise modeling of demand by customer using the customer's seasonal sales history. Especially useful was the ability to see historical sales lift during promotional periods. The forecast accuracy cut overproduction and improved customer service, which both reduced costs and improved the top line.

- Automatic three-way matching saved $2 million in goods and services purchases. An ERP system we implemented had the standard financial package that required vendor invoice amounts to be the same as the purchase order and for the same quantity that was physically received. With non-integrated systems these three activities are normally in different software applications: purchase orders in one, receipts in another, and invoice processing in a third.

- A reduction of $1.5 million in manufacturing waste via implementation of standard costing and variance analysis. Prior to ERP, the company had used systems that only man-aged the production process itself – a shop floor system. Costs were not part of this system, so every day the account-ing staff would manually extract from the shop floor system actual production quantities and actual raw materials used, plug the data into a spreadsheet, add in labor and overhead costs, and calculate total manufacturing costs. The ERP sys-tem automatically took the shop floor data, compared it to standards (or budgeted costs) and displayed cost variance reports identifying waste.

From Chapter 1: "I once heard a CEO say his company's SAP implementation was like 'a fish riding a bicycle'."

*Illustration by John P. Sczerba*

www.ingramcontent.com/pod-product-compliance
Lightning Source LLC
Chambersburg PA
CBHW020200200326
41521CB00005BA/201